Selected Titles in This Series

695 **Vitaly Bergelson and Randall McCutcheon,** An ergodic IP polynomial Szemerédi theorem, 2000

694 **Alberto Bressan, Graziano Crasta, and Benedetto Piccoli,** Well-posedness of the Cauchy problem for $n \times n$ systems of conservation laws, 2000

693 **Doug Pickrell,** Invariant measures for unitary groups associated to Kac-Moody Lie algebras, 2000

692 **Mara D. Neusel,** Inverse invariant theory and Steenrod operations, 2000

691 **Bruce Hughes and Stratos Prassidis,** Control and relaxation over the circle, 2000

690 **Robert Rumely, Chi Fong Lau, and Robert Varley,** Existence of the sectional capacity, 2000

689 **M. A. Dickmann and F. Miraglia,** Special groups: Boolean-theoretic methods in the theory of quadratic forms, 2000

688 **Piotr Hajłasz and Pekka Koskela,** Sobolev met Poincaré, 2000

687 **Guy David and Stephen Semmes,** Uniform rectifiability and quasiminimizing sets of arbitrary codimension, 2000

686 **L. Gaunce Lewis, Jr.,** Splitting theorems for certain equivariant spectra, 2000

685 **Jean-Luc Joly, Guy Metivier, and Jeffrey Rauch,** Caustics for dissipative semilinear oscillations, 2000

684 **Harvey I. Blau, Bangteng Xu, Z. Arad, E. Fisman, V. Miloslavsky, and M. Muzychuk,** Homogeneous integral table algebras of degree three: A trilogy, 2000

683 **Serge Bouc,** Non-additive exact functors and tensor induction for Mackey functors, 2000

682 **Martin Majewski,** ational homotopical models and uniqueness, 2000

681 **David P. Blecher, Paul S. Muhly, and Vern I. Paulsen,** Categories of operator modules (Morita equivalence and projective modules, 2000

680 **Joachim Zacharias,** Continuous tensor products and Arveson's spectral C^*-algebras, 2000

679 **Y. A. Abramovich and A. K. Kitover,** Inverses of disjointness preserving operators, 2000

678 **Wilhelm Stannat,** The theory of generalized Dirichlet forms and its applications in analysis and stochastics, 1999

677 **Volodymyr V. Lyubashenko,** Squared Hopf algebras, 1999

676 **S. Strelitz,** Asymptotics for solutions of linear differential equations having turning points with applications, 1999

675 **Michael B. Marcus and Jay Rosen,** Renormalized self-intersection local times and Wick power chaos processes, 1999

674 **R. Lawther and D. M. Testerman,** A_1 subgroups of exceptional algebraic groups, 1999

673 **John Lott,** Diffeomorphisms and noncommutative analytic torsion, 1999

672 **Yael Karshon,** Periodic Hamiltonian flows on four dimensional manifolds, 1999

671 **Andrzej Rosłanowski and Saharon Shelah,** Norms on possibilities I: Forcing with trees and creatures, 1999

670 **Steve Jackson,** A computation of δ_5^1, 1999

669 **Seán Keel and James McKernan,** Rational curves on quasi-projective surfaces, 1999

668 **E. N. Dancer and P. Poláčik,** Realization of vector fields and dynamics of spatially homogeneous parabolic equations, 1999

667 **Ethan Akin,** Simplicial dynamical systems, 1999

666 **Mark Hovey and Neil P. Strickland,** Morava K-theories and localisation, 1999

665 **George Lawrence Ashline,** The defect relation of meromorphic maps on parabolic manifolds, 1999

(*Continued in the back of this publication*)

An Ergodic IP Polynomial Szemerédi Theorem

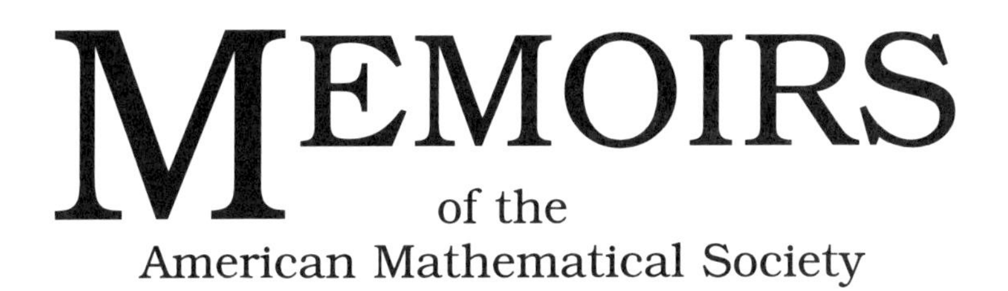

Number 695

An Ergodic IP Polynomial Szemerédi Theorem

Vitaly Bergelson
Randall McCutcheon

July 2000 • Volume 146 • Number 695 (fourth of 5 numbers) • ISSN 0065-9266

American Mathematical Society
Providence, Rhode Island

2000 *Mathematics Subject Classification.*
Primary 28D05; Secondary 05A17, 05D10, 05A10, 11B05, 11B83.

Library of Congress Cataloging-in-Publication Data

Bergelson, V. (Vitaly), 1950–
An ergodic IP polynomial Szemerédi theorem /Vitaly Bergelson, Randall McCutcheon
p. cm. — (Memoirs of the American Mathematical Society, ISSN 0065 9266 , no. 695)
Includes bibliographical references and index.
ISBN 0-8218-2657-3
1. Measure-preserving transformations. 2. Ramsey theory. I. McCutcheon, Randall, 1965–
II. Title. III. Series.
QA3 .A57 no. 695+
[QA313]
510 s—dc21
[515′.42] 00-036258

Memoirs of the American Mathematical Society

This journal is devoted entirely to research in pure and applied mathematics.

Subscription information. The 2000 subscription begins with volume 143 and consists of six mailings, each containing one or more numbers. Subscription prices for 2000 are $466 list, $419 institutional member. A late charge of 10% of the subscription price will be imposed on orders received from nonmembers after January 1 of the subscription year. Subscribers outside the United States and India must pay a postage surcharge of $30; subscribers in India must pay a postage surcharge of $43. Expedited delivery to destinations in North America $35; elsewhere $130. Each number may be ordered separately; *please specify number* when ordering an individual number. For prices and titles of recently released numbers, see the New Publications sections of the *Notices of the American Mathematical Society.*

Back number information. For back issues see the *AMS Catalog of Publications.*

Subscriptions and orders should be addressed to the American Mathematical Society, P. O. Box 5904, Boston, MA 02206-5904. *All orders must be accompanied by payment.* Other correspondence should be addressed to Box 6248, Providence, RI 02940-6248.

Memoirs of the American Mathematical Society is published bimonthly (each volume consisting usually of more than one number) by the American Mathematical Society at 201 Charles Street, Providence, RI 02904-2294. Periodicals postage paid at Providence, RI. Postmaster: Send address changes to Memoirs, American Mathematical Society, P. O. Box 6248, Providence, RI 02940-6248.

This publication is indexed in *Science Citation Index®*, *SciSearch®*, *Research Alert®*, *CompuMath Citation Index®*, *Current Contents®/Physical, Chemical & Earth Sciences.*
Printed in the United States of America.

∞ The paper used in this book is acid-free and falls within the guidelines established to ensure permanence and durability.
Visit the AMS home page at URL: `http://www.ams.org/`

10 9 8 7 6 5 4 3 2 1 05 04 03 02 01 00

CONTENTS

Abstract

We prove a polynomial multiple recurrence theorem for finitely many commuting measure preserving transformations of a probability space, extending a polynomial Szemerédi theorem appearing in [BL1]. The linear case is a consequence of an ergodic IP-Szemerédi theorem of Furstenberg and Katznelson ([FK2]). Several applications to the fine structure of recurrence in ergodic theory are given, some of which involve weakly mixing systems, for which we also prove a multiparameter weakly mixing polynomial ergodic theorem. The techniques and apparatus employed include a polynomialization of an IP structure theory developed in [FK2], an extension of Hindman's theorem due to Milliken and Taylor ([M], [T]), a polynomial version of the Hales-Jewett coloring theorem ([BL2]), and a theorem concerning limits of polynomially generated IP-systems of unitary operators ([BFM]).

1991 Mathematics Subject Classification: Primary: 28D05
Secondary: 05A17, 05D10, 11B05, 11B83.

Keywords and phrases: Ergodic Ramsey theory; Furstenberg correspondence principle; IP-sets; Hindman's theorem; Mild mixing; Multiple recurrence; Polynomial Hales-Jewett theorem; Polynomial Szemerédi theorem; Weak mixing.

INTRODUCTION

A celebrated theorem of Szemerédi ([SZ]) states that if a set $S \subset \mathbf{Z}$ has positive upper density

$$\overline{d}(S) = \limsup_{N\to\infty} \frac{|S \cap \{-N, \cdots, N\}|}{2N+1} > 0,$$

then S contains arbitrarily long arithmetic progressions.

Soon after Szemerédi's theorem appeared, H. Furstenberg gave in [F1] a new, ergodic theoretical proof of Szemerédi's theorem by deducing it from a far-reaching extension of the classical Poincaré recurrence theorem. A short time later, in [FK1], Furstenberg and Katznelson proved the following *multiple recurrence theorem*:

Theorem 0.1 Suppose that $(X, \mathcal{B}, \mu)$ is a probability space and that $T_1, \cdots, T_k$ are commuting measure-preserving transformations of X. For every $A \in \mathcal{B}$ with $\mu(A) > 0$ we have

$$\liminf_{N\to\infty} \frac{1}{N} \sum_{n=0}^{N-1} \mu(T_1^{-n} A \cap \cdots \cap T_k^{-n} A) > 0.$$

As a corollary of this result they obtained a multi-dimensional generalization of Szemerédi's theorem for which there is as yet no non-ergodic proof. We will now formulate this result. The *Banach upper density* of a set $S \subset \mathbf{Z}^k$ is defined to be

$$d^*(S) = \sup_{\{\Pi_n\}_{n\in\mathbf{N}}} \limsup_{n\to\infty} \frac{|S \cap \Pi_n|}{|\Pi_n|},$$

where the supremum goes over all sequences of parallelepipeds

$$\Pi_n = [a_n^{(1)}, b_n^{(1)}] \times \cdots \times [a_n^{(k)}, b_n^{(k)}] \subset \mathbf{Z}^k, n \in \mathbf{N},$$

with $b_n^{(i)} - a_n^{(i)} \to \infty$, $1 \le i \le k$.

Corollary 0.2 ([FK1], Theorem B) Suppose that $S \subset \mathbf{Z}^k$ with $d^*(S) > 0$ and that $F \subset \mathbf{Z}^k$ is a finite configuration. There exists a positive integer n and a vector $u \in \mathbf{Z}^k$ such that $u + nF = \{u + nx : x \in F\} \subset S$.

The derivation of combinatorial results such as Corollary 0.2 from recurrence results hinges on a general correspondence principle due Furstenberg.

Furstenberg's Correspondence Principle Given $E \subset \mathbf{Z}^k$ with $d^*(E) > 0$ there is a probability space $(X, \mathcal{B}, \mu)$ and k commuting invertible measure preserving transformations $T_1, T_2, \cdots, T_k$ of X such that for any $\mathbf{n}_1, \mathbf{n}_2, \cdots, \mathbf{n}_l \in \mathbf{Z}^k$ one has

$$d^*(E \cap (E - \mathbf{n}_1) \cap (E - \mathbf{n}_2) \cap \cdots \cap (E - \mathbf{n}_l)) \ge \mu(A \cap T_{\mathbf{n}_1} \cap \cdots \cap T_{\mathbf{n}_l} A),$$

Received by the editor February 5, 1998.

The authors acknowledge the support of the NSF under grant DMS-9706057 and a postodoctoral fellowship administered by the University of Maryland, respectively.

where for $\mathbf{n} = (n_1, \cdots, n_k)$, $T_{\mathbf{n}} = T_1^{n_1} \cdots T_k^{n_k}$.

[F1] also contained an ergodic proof of a non-linear number theoretical result which had been proved independently by Sárközy ([S]) and by Conze: If $S \subset \mathbf{Z}$ has positive upper Banach density, then there exist $x, y \in S$ with $x - y = n^2$ for some $n \in \mathbf{N}$. This result is readily extendable to more general polynomials:

Theorem 0.3 ([F2], [KM], [S]) Suppose that $p(n) \in \mathbf{Q}[n]$ is a polynomial taking on integer values on the integers and satisfying $p(0) = 0$. Then if $S \subset \mathbf{Z}$ has positive upper density $\overline{d}(S) > 0$, there exist $n \in \mathbf{N}$ and $x, y \in S$ with $x - y = p(n) \neq 0$.

In [F2] this theorem is proved via a *polynomial recurrence theorem*. Namely, it is shown that for any invertible measure preserving system $(X, \mathcal{B}, \mu, T)$ with $\mu(X) = 1$, any $p(n) \in \mathbf{Q}[n]$ with $p(\mathbf{Z}) \subset \mathbf{Z}$ and $p(0) = 0$, and any $A \in \mathcal{B}$, $\mu(A) > 0$, there exists $n \in \mathbf{N}$ with $p(n) \neq 0$ such that $\mu(A \cap T^{-p(n)}A) > 0$. What is actually proved is stronger, namely that

$$\lim_{N\to\infty} \frac{1}{N} \sum_{n=1}^{N} \mu(A \cap T^{-p(n)}A) > 0,$$

which follows from the study of certain Césaro averages of polynomial powers of unitary operators on a Hilbert space, utilizing the spectral theorem. (See [B1] for a treatment which avoids the use of the spectral theorem.)

More recently, Bergelson and Leibman have proved a *polynomial multiple recurrence theorem*. Notice that by taking $k = t = 1$ in Theorem 0.4 below and applying Furstenberg's correspondence principle one gets Theorem 0.3. Theorem 0.1 is a special case as well, corresponding to the case of linear polynomials.

Theorem 0.4 ([BL1]) Suppose that $(X, \mathcal{B}, \mu)$ is a probability space and that $T_1, \cdots, T_t$ are commuting invertible measure preserving transformations of X. If $\{p_{i,j}(n) : 1 \leq i \leq k, 1 \leq j \leq t\} \subset \mathbf{Q}[n]$ satisfy $p_{i,j}(\mathbf{Z}) \subset \mathbf{Z}$ and $p_{i,j}(0) = 0$, then for every $A \in \mathcal{B}$ with $\mu(A) > 0$ one has

$$\liminf_{N\to\infty} \frac{1}{N} \sum_{n=1}^{N} \mu\Big(\bigcap_{i=1}^{k} \Big(\prod_{j=1}^{t} T_j^{p_{i,j}(n)} \Big)^{-1} A \Big) > 0.$$

As a consequence of this theorem and Furstenberg's correspondence principle, one obtains a "multidimensional polynomial Szemerédi theorem":

Corollary 0.5 ([BL1]) Suppose that $r, l \in \mathbf{N}$ and $P : \mathbf{Z}^r \to \mathbf{Z}^l$ is a polynomial mapping which satisfies $P(0) = 0$. If $F \subset \mathbf{Z}^r$ is any finite configuration then for any subset $S \subset \mathbf{Z}^l$ of positive upper Banach density there exist $n \in \mathbf{N}$ and $u \in \mathbf{Z}^l$ such that $u + P(nF) = \{u + P(nx) : x \in F\} \subset S$.

In [BFM], a different type of refinement of Theorem 0.3 was obtained as a corollary of a general theorem concerning *weak IP-convergence* of certain polynomially-generated sequences of unitary operators on a Hilbert space. In order to formulate the primary results of [BFM], as well as those of this paper, we will need to introduce some definitions and notation.

Given an infinite sequence $G = \{g_i : i \in \mathbf{N}\} \subset \mathbf{Z}$, the *IP-set* generated by G is the set

$$\Gamma = \{g_{i_1} + g_{i_2} + \cdots + g_{i_k} : i_1 < i_2 < \cdots < i_k, k \in \mathbf{N}\}$$

of all finite sums of elements with distinct indices from G. IP-sets in $\mathbf{Z}^k$, $k > 1$, are defined similarly (for a sequence $(g_i) \subset \mathbf{Z}^k$). IP-sets are generally expressed as indexed sequences. Let $\mathcal{F}$ denote the family of finite subsets of $\mathbf{N}$. An IP-set, then, may be viewed as a sequence indexed by $\mathcal{F}$, $(n_\alpha)_{\alpha\in\mathcal{F}}$, which satisfies $n_{\alpha\cup\beta} = n_\alpha + n_\beta$ for $\alpha \cap \beta = \emptyset$. (See also Section 1, where these ideas are developed more fully.)

The following *IP polynomial recurrence theorem* is a special case of a result from [BFM]. It may be seen to be a multi-operator, IP-set generalization of Theorem 0.3.

Theorem 0.6 Suppose that we have an IP-set $\Gamma \subset \mathbf{Z}$ and r commuting invertible measure preserving transformations $T_1, \cdots, T_r$ of a probability space $(X, \mathcal{B}, \mu)$. Suppose we are given polynomials $p_i(n) \in \mathbf{Z}[n]$ with $p_i(0) = 0$, $1 \le i \le r$. Then for every $A \in \mathcal{B}$ with $\mu(A) > 0$ there exists $n \in \Gamma$ such that

$$\mu\Big(A \cap \Big(\prod_{i=1}^{r} T_i^{p_i(n)}\Big)^{-1} A\Big) > 0.$$

Although Theorem 0.6 involves multiple transformations, note that it is not a multiple recurrence theorem, as it guarantees only a single return of the set A to itself. Theorem 0.4, on the other hand, is not an IP recurrence theorem. Finally, Furstenberg and Katznelson have a (linear) *IP multiple recurrence theorem* ([FK1]). It is our purpose in this paper to extend all of these previous results by proving an *IP polynomial multiple recurrence theorem.* This is accomplished via our main theorem, Theorem 1.3, with one limitation we shall address in the next paragraph. Theorem 0.4 is completely generalized with a uniform, multiparameter version given in Theorem 6.13. Theorem 0.6 is generalized by the following.

Theorem 0.7 Suppose that we have an IP-set $\Gamma \subset \mathbf{Z}$ and r commuting invertible measure preserving transformations $T_1, \cdots, T_r$ of a probability space $(X, \mathcal{B}, \mu)$. Suppose $t \in \mathbf{N}$ and that $p_{i,j}(n) \in \mathbf{Q}[n]$ with $p_{i,j}(\mathbf{Z}) \subset \mathbf{Z}$ and $p_{i,j}(0) = 0$, $1 \le i \le r$, $1 \le j \le t$. Then for every $A \in \mathcal{B}$ with $\mu(A) > 0$ there exists $n \in \Gamma$ such that

$$\mu\Big(\bigcap_{j=1}^{t} \Big(\prod_{i=1}^{r} T_i^{p_{i,j}(n)}\Big)^{-1} A\Big) > 0.$$

The limitation we referred to earlier concerns our ability to fully generalize the IP multiple recurrence theorem of Furstenberg and Katznelson (there are some interesting open problems related to potential fuller generalizations, as we shall discuss in Section 8 below). The linear case of Theorem 0.7, that is, the case in which all of the polynomials appearing there are of first degree, as well as the linear cases of the other IP polynomial multiple recurrence theorems in this paper, including Theorem 1.3, follows from Theorem A in [FK1]. This theorem is quite a bit more powerful, however, in that it allows one to guarantee multiple recurrence along countably generated IP-systems of commuting measure preserving transformations. In the non-linear case, we do not see at this time how one should deal with such general systems. (For a fuller discussion of these matters, see Section 8.) Therefore in this paper we will be proving results which are at once stonger and weaker than those of [FK1]. They are stronger in that they are non-linear, and weaker in that we restrict ourselves to a special class of IP-systems–namely those obtained by taking finite families of commuting measure preserving transformations to various polynomial powers.

The following is the most natural combinatorial consequence of Theorem 0.7.

Corollary 0.8 Suppose that we are given an IP-set $\Gamma \subset \mathbf{Z}$ and a subset $E \subset \mathbf{Z}^r$ having positive upper Banach density. Suppose that $t \in \mathbf{N}$ and $p_{i,j}(n) \in \mathbf{Q}[n]$ with $p_{i,j}(\mathbf{Z}) \subset \mathbf{Z}$ and $p_{i,j}(0) = 0$, $1 \leq i \leq r$, $1 \leq j \leq t$. Define vector valued functions

$$v_j(n) = \big(p_{1,j}(n), \cdots, p_{r,j}(n)\big), \quad 1 \leq j \leq t. \tag{0.1}$$

Under these conditions, there exists $u \in \mathbf{Z}^r$ and $n \in \Gamma$ such that $u + v_j(n) \in E$, $1 \leq j \leq t$.

One can show that Corollary 0.8 may be used to give a refinement of Corollary 0.5. Namely, one has that if $r, l \in \mathbf{N}$, $P : \mathbf{Z}^r \to \mathbf{Z}^l$ is a polynomial mapping which satisfies $P(0) = 0$, and $F \subset \mathbf{Z}^r$ is any finite configuration, then for any subset $S \subset \mathbf{Z}^l$ of positive upper Banach density and any IP-set $\Gamma \subset \mathbf{Z}$, there exist $n \in \Gamma$ and $u \in \mathbf{Z}^l$ such that $u + P(nF) \subset S$. (Confer with the last section of [BL1].)

Given a probability space $(X, \mathcal{B}, \mu)$, a finite set of commuting, invertible measure preserving transformations $T_1, \cdots, T_k$ of X, a set $A \in \mathcal{B}$, $\mu(A) > 0$, and polynomials $p_{i,j}(n) \in \mathbf{Q}[n]$ with $p_{i,j}(\mathbf{Z}) \subset \mathbf{Z}$ and $p_{i,j}(0) = 0$, $1 \leq i \leq k$, $1 \leq j \leq t$, let

$$R_A = \Big\{ n \in \mathbf{Z} : \ \mu\Big(\bigcap_{j=1}^{t} \Big(\prod_{i=1}^{k} T_i^{p_{i,j}(n)} \Big)^{-1} A \Big) > 0 \Big\}. \tag{0.2}$$

It follows from Theorem 0.4 that the set R_A has positive lower density

$$\underline{d}(R_A) = \liminf_{N \to \infty} \frac{|R_A \cap \{-N, \cdots, N\}|}{2N+1} > 0$$

in $\mathbf{Z}$. We shall explain now in terms of the largeness of the set R_A why Theorem 0.7 gives more.

A subset $E \subset \mathbf{Z}^k$, $k \geq 1$, is called an *IP*-set* if for any IP-set $S \subset \mathbf{Z}^k$ one has $E \cap S \neq \emptyset$. It is not hard to show (see Proposition 6.11) that any IP*-set is *relatively dense*, that is, *syndetic*. (We remark that the notions of syndeticity, IP-set, and IP*-set have meaning in an arbitrary (semi)group. We will concern ourselves here only with $\mathbf{Z}^k$. A set $E \subset \mathbf{Z}^k$ is said to be syndetic in $\mathbf{Z}^k$ if there is some finite set $F \subset \mathbf{Z}^k$ such that $E + F = \{x + y : \ x \in E, y \in F\} = \mathbf{Z}^k$. In $\mathbf{Z}$, therefore, the syndetic sets are those which do not have arbitrarily large gaps. In particular, syndetic sets have positive lower density.) It is easy to construct sets which have positive lower density but are not syndetic, or sets which are syndetic but not IP*. (An example of the latter in $\mathbf{Z}$ is the set of odd integers. As a curiosity, we mention that one can even construct syndetic IP-sets which are not IP*; see Example 7.9.) Hence we see that the class of IP*-sets is much more exclusive than the class of syndetic sets.

Furthermore, a consequence of Hindman's theorem ([H], see Proposition 2.13 below) is that the intersection of any two IP*-sets in $\mathbf{Z}$ (or $\mathbf{Z}^k$) is again an IP*-set. This finite intersection property gives us a yet further sense that IP*-sets are quite "large". In this sense, then, Corollary 0.8 guarantees "many" (namely "IP*-many") n which may act as the parameter for the special type of polynomial configurations under consideration. In other words, it guarantees, in the notation used there, that the set

$$R = \{n : u + v_j(n) \in E, \ 1 \leq j \leq t\}$$

is an IP*-set. This is an improvement on Theorem B of [BL1], which gives only positive lower density of the set R, and on Theorem 0.3 of [BM], which gives syndeticity of the set R in the case $r = 1$ (i.e. for $E \subset \mathbf{Z}$).

IP*-sets are significantly more combinatorially rich than syndetic sets. One can show, for example, that if E is an IP*-set in $\mathbf{Z}^k$ then for any IP-sets $S_1, S_2, \cdots, S_k$ in $\mathbf{Z}$ one can find arbitrarily large finite sets $I_i \subset S_i$, $1 \leq i \leq k$, such that $I_1 \times I_2 \times \cdots \times I_k \subset E$. One might ask whether one can strengthen this fact by requiring each set I_i to be an infinite subset of S_i which is also an IP-set. For a general IP*-set in $\mathbf{Z}^k$, the answer is no (see [BH]). However, for the set R_A of (0.2), which by Theorem 0.7 is already an IP*-set, the answer turns out to be yes. Namely, for any IP-sets $S_1, S_2, \cdots, S_k$ in $\mathbf{Z}$ one can find IP-sets $I_i \subset S_i$, $1 \leq i \leq k$, such that $I_1 \times I_2 \times \cdots \times I_k \subset R_A$ (this is demonstrated in Section 6).

This provides an impetus for studying a class of sets having this property. Such is the case for a class of sets we call *enhanced IP*-sets*, or simply *E-IP*-sets*. Most of the sets we show to be IP* are actually E-IP*. As a matter of fact, they have an even stronger property which we call the PE-IP* property, PE standing for *polynomially enhanced.* We will give a precise definition of both the E-IP* and PE-IP* properties in Section 6. For now, we mention only that if a set A is a PE-IP*-set in $\mathbf{Z}^k$, $S_1, S_2, \cdots, S_m$ are IP-sets in $\mathbf{Z}$, and $P : \mathbf{Z}^m \to \mathbf{Z}^k$ is a polynomial mapping with $P(0) = 0$ then there exist IP-sets $I_i \subset S_i$, $1 \leq i \leq m$, such that $P(I_1 \times I_2 \times \cdots \times I_m) \subset A$.

As we noted, Theorem 0.7 asserts that the set R_A of (0.2) is an IP*-set in $\mathbf{Z}$. In fact, it is a PE-IP*-set and moreover we have a natural extension of this fact in $\mathbf{Z}^k$ which is also a consequence of our main theorem, Theorem 1.3.

Theorem 0.9 Suppose we are given r commuting invertible measure preserving transformations $T_1, \cdots, T_r$ of a probability space $(X, \mathcal{B}, \mu)$. Let $k, t \in \mathbf{N}$, and suppose that $p_{i,j}(n_1, \cdots, n_k) \in \mathbf{Q}[n_1, \cdots, n_k]$ with $p_{i,j}(\mathbf{Z}^k) \subset \mathbf{Z}$ and $p_{i,j}(0, \cdots, 0) = 0$, $1 \leq i \leq r$, $1 \leq j \leq t$. Then for every $A \in \mathcal{B}$ with $\mu(A) > 0$ the set

$$R_A = \left\{(n_1, \cdots, n_k) \in \mathbf{Z}^k : \mu\left(\bigcap_{j=1}^{t} \left(\prod_{i=1}^{r} T_i^{p_{i,j}(n_1,\cdots,n_k)}\right)^{-1} A\right) > 0\right\}$$

is a PE-IP*-set in $\mathbf{Z}^k$.

We remark that a version of the foregoing theorem holds for non-invertible systems as well (see Theorem 7.12 below).

Because of the strength of the PE-IP* property, Theorem 0.9 is actually equivalent to its own linear version. Namely, it is an easy matter to get Theorem 0.9 in general provided one has it for the special case in which all of the polynomials $p_{i,j}$ are linear.

The case $t = 2$ of Theorem 0.9 was proved in [BFM], at least for the IP* property. It would not be difficult to get from there to establishing the PE-IP* property for this, the single recurrence, case.

In order to give the reader a feel for the multifareousness of Theorem 0.9 we shall give now a few examples of applications of the result and/or its proof to combinatorics and ergodic theory. A more comprehensive examination of combinatorial consequences of this theorem and of our main theorem, Theorem 1.3, which is an even stronger multiple recurrence result, is deferred to Section 7. We start with some combinatorics.

Being a statement about "diagonals" of polynomials of many variables (this will presently be made clearer), Theorem 0.9 enlarges our knowledge about the types of configurations which one is always promised to find in any set of positive density in $\mathbf{Z}^k$, as the following corollary shows.

Theorem 0.10 Suppose $r, l \in \mathbf{N}$ and let $P : \mathbf{Z}^r \to \mathbf{Z}^l$ be a polynomial mapping satisfying $P(0) = 0$. Let $F \subset \mathbf{Z}^r$ be a finite set, let $S \subset \mathbf{Z}^l$ be a set of positive upper Banach density, and let $(n_\alpha^{(i)})_{\alpha \in \mathcal{F}}$ be arbitrary IP-sets in $\mathbf{Z}$, $1 < i \leq r$. Then for some $u \in \mathbf{Z}^l$ and $\alpha \in \mathcal{F}$ one has:

$$\big\{u + P(n_\alpha^{(1)}x_1, n_\alpha^{(2)}x_2, \cdots, n_\alpha^{(r)}x_r) : (x_1, x_2, \cdots, x_r) \in F\big\} \subset S.$$

With a little effort, one may show that Theorem 0.10 generalizes Corollary 0.8 (and therefore Corollary 0.5 as well). As a particular application of Theorem 0.10, one has that for any set of positive upper Banach density $E \subset \mathbf{Z}$ and for any IP-sets $(n_\alpha^{(i)})_{\alpha \in \mathcal{F}}$, $1 \leq i \leq k$ there exist $x \in E$ and $\alpha \in \mathcal{F}$ such that

$$\big\{x, x + n_\alpha^{(1)}, x + n_\alpha^{(1)}n_\alpha^{(2)}, \cdots, x + n_\alpha^{(1)}n_\alpha^{(2)} \cdots n_\alpha^{(k)}\big\} \subset E.$$

The following application pertains to partition Ramsey theory rather than to density Ramsey theory (for a more general result see Theorem 7.4–cf. also [BM], Theorem 0.4).

Theorem 0.11 Let $t \in \mathbf{N}$ and let $p_i(x, y), q_i(x, y) \in \mathbf{Z}[x, y]$ with $p_i(0, 0) = q_i(0, 0) = 0$, $1 \leq i \leq t$. Suppose that $s \in \mathbf{N}$ and that $\mathbf{Z}^2 = \bigcup_{i=1}^s C_i$ is a partition of $\mathbf{Z}^2$ into s cells. Then there exists some $L \in \mathbf{N}$ and some $\epsilon > 0$ having the property that in any rectangle $B = [M_1, N_1] \times [M_2, N_2] \subset \mathbf{Z}^2$ with $\min\{N_1 - M_1, N_2 - M_2\} \geq L$ there exists i with $1 \leq i \leq s$ and $(n, m) \in C_i \cap B$ such that

$$d^*\Big(C_i \cap \big(C_i - (p_1(n, m), q_1(n, m))\big) \cap \cdots \cap \big(C_i - \big(p_t(n, m), q_t(n, m)\big)\big)\Big) > \epsilon.$$

In particular, the system of polynomial equations

$$\begin{aligned}
(x_0, y_0) &= (n, m),\\
(x_2, y_2) - (x_1, y_1) &= \big(p_1(n, m), q_1(n, m)\big)\\
(x_3, y_3) - (x_1, y_1) &= \big(p_2(n, m), q_2(n, m)\big)\\
&\vdots\\
(x_{t+1}, y_{t+1}) - (x_1, y_1) &= \big(p_t(n, m), q_t(n, m)\big)
\end{aligned}$$

has monochromatic solutions $\big\{(x_0, y_0), (x_1, y_1), \cdots, (x_{t+1}, y_{t+1})\big\}$ with (n, m) choosable from any large enough rectangle in $\mathbf{Z}^2$.

We pass now to a description of some applications pertaining to ergodic theory. The first of these has to do with recurrence properties of weak mixing. Recall that an invertible measure preserving transformation T of a probability space $(X, \mathcal{B}, \mu)$ is called *weakly mixing* if the induced unitary operator U_T defined by $(U_T f)(x) = f(Tx)$, $f \in L^2(X, \mathcal{B}, \mu)$ has no non-constant eigenfunctions. The notion of weak mixing is in a sense complementary to that of being isomorphic to a rotation on a compact abelian group and can be characterized in a variety of ways. See, for example, [Ha], where among other things weak mixing is characterized by the quality of the martini cocktail. Also, the notion of (relative) weak mixing is an important ingredient in Furstenberg's structure theorem for measure preserving systems, which was used in his proof of Szemerédi's theorem ([F1]).

A feature of weak mixing which plays a crucial role in proving results like Theorem 0.4 is that it implies weak mixing of higher orders. One can prove, for example, the following theorem:

Theorem 0.12 ([B1]). If $(X, \mathcal{B}, \mu, T)$ is an invertible weakly mixing measure preserving system, $A_0, A_1, \cdots, A_k \in B$ and $p_1(t), \cdots, p_k(t) \in \mathbf{Q}[t]$ are non-constant polynomials with $p_i(\mathbf{Z}) \subset \mathbf{Z}$ and such that $p_i(t) - p_j(t)$ is not constant for $1 \leq i \neq j \leq k$, then for any $\epsilon > 0$ the set

$$S = \{n : \left|\mu(A_0 \cap T^{p_1(n)} A_1 \cap \cdots \cap T^{p_k(n)} A_k) - \mu(A_0)\mu(A_1) \cdots \mu(A_k)\right| < \epsilon\}$$

has *uniform* density one, that is

$$\lim_{N-M\to\infty} \frac{\left|S \cap \{M, M+1, \cdots, N-1\}\right|}{N-M} = 1.$$

It is a peculiar fact that although weakly mixing transformations admit no eigenfunctions, they may allow for so-called *rigid* functions, namely functions which for some sequence $n_i \to \infty$ satisfy $U_T^{n_i} f \to f$ in L^2 norm. One can show that the "typical" weak mixing system $(X, \mathcal{B}, \mu, T)$ has the property that all functions in $L^2(X, \mathcal{B}, \mu)$ are rigid. We shall call such systems rigid. It follows that while being of uniform density 1, the set S in the formulation of Theorem 0.12 is not necessarily an IP*-set. For example, if $(X, \mathcal{B}, \mu, T)$ is a rigid weakly mixing system and if for some $A \in \mathcal{B}$ with $0 < \mu(A) < 1$ one has $U_T^{n_i} 1_A \to 1_A$ then it is not hard to see that for any small enough $\epsilon > 0$ the set

$$\{n : \left|\mu(A \cap T^n A) - \mu^2(A)\right| > \epsilon\}$$

contains an IP-set (generated by a subsequence of $(n_i)_{i=1}^{\infty}$). It follows that the set

$$\{n : \left|\mu(A \cap T^n A) - \mu^2(A)\right| < \epsilon\}$$

is not an IP*-set.

This should be compared with the easily provable fact (see for example [B2], Section 5) that for any measure preserving system $(X, \mathcal{B}, \mu, T)$, any $A \in \mathcal{B}$ and any $\epsilon > 0$, the set

$$\{n : \mu(A \cap T^n A) \geq \mu(A)^2 - \epsilon\}$$

is an IP*-set. Let us call a set $E \subset \mathbf{Z}$ an *IP$^*_+$-set* if E is a shift of an IP*-set. *E-IP$^*_+$-sets* and *PE-IP$^*_+$-sets* are defined similarly. IP$^*_+$-sets are, so to say, affine versions of IP*-sets and share with them many of their features of largeness and combinatorial richness. In particular, IP$^*_+$-sets form a significantly smaller family than that of syndetic sets. Indeed, in Example 7.5 we give a construction of a syndetic set which is not an IP$^*_+$-set . (Example 7.8 is even stronger.)

One can easily show that for any $\alpha, \beta > 0$ there exists a positive constant c such if $(X, \mathcal{B}, \mu, T)$ is an ergodic system (i.e. there are no non-trivial invariant sets) and $A, B \in \mathcal{B}$ with $\mu(A) > \alpha$ and $\mu(B) > \beta$ then the set $\{n : \mu(A \cap T^n B) \geq c\}$ is an IP$^*_+$-set . (One can take $c = \alpha^2\beta^2$; we do not know whether or not this can be improved.) The following theorem gives an analogue of this fact for polynomial recurrence of so-called *totally weakly mixing* systems (that is, systems $\big(X, \mathcal{B}, \mu, \{T_{\mathbf{n}}\}_{\mathbf{n}\in\mathbf{Z}^r}\big)$, where $T_{\mathbf{n}}$ is weakly mixing for all $\mathbf{n} \in \mathbf{Z}^k \setminus \{\mathbf{0}\}$).

Theorem 0.13 Suppose that $(X, \mathcal{B}, \mu)$ is a probability space and $\{T_{\mathbf{n}}\}_{\mathbf{n}\in\mathbf{Z}^r}$ is a totally weakly mixing measure preserving $\mathbf{Z}^r$-action generated by $T_1, \cdots, T_r$. Suppose that $t \in \mathbf{N}$ and $p_{i,j}(x_1, \cdots, x_k) \in \mathbf{Q}[x_1, \cdots, x_k]$ with $p_{i,j}(\mathbf{Z}^k) \subset \mathbf{Z}$, $1 \le i \le r$, $1 \le j \le t$ such that for any $1 \le j_1 \ne j_2 \le t$, the mappings

$$(l_1, \cdots, l_k) \to \big(p_{1,j_1}(l_1, \cdots, l_k), \cdots, p_{r,j_1}(l_1, \cdots, l_k)\big)$$

and

$$(l_1, \cdots, l_k) \to \big(p_{1,j_1}(l_1, \cdots, l_k) - p_{1,j_2}(l_1, \cdots, l_k), \cdots, p_{r,j_1}(l_1, \cdots, l_k) - p_{r,j_2}(l_1, \cdots, l_k)\big)$$

are not constant. Suppose that $A_0, A_1, A_2, \cdots, A_t \in \mathcal{B}$ with $\mu(A_i) > 0$, $0 \le i \le t$. Then the set

$$R_{A_0, A_1, \cdots, A_r} = \left\{(n_1, \cdots, n_k) \in \mathbf{Z}^k : \mu\Big(A_0 \cap \bigcap_{j=1}^{t} \Big(\prod_{i=1}^{r} T_i^{p_{i,j}(n_1, \cdots, n_k)}\Big)^{-1} A_i\Big) > 0\right\}$$

is a PE-IP$^*_+$-set.

In the hierarchy of mixing there is a notion, namely that of *mild mixing*, which was introduced by P. Walters in [W] and later rediscovered and given an "IP* flavor" by Furstenberg and Weiss (see for example [FW] or [F2], section 9.4) which falls between weak mixing and *strong mixing* (a measure preserving transformation T on a probability space $(X, \mathcal{B}, \mu)$ is strongly mixing if for every $A, B \in \mathcal{B}$ one has $\lim_{n\to\infty} \mu(A \cap T^n B) = \mu(A)\mu(B)$) and which is characterized by the absense of rigid functions. Namely, a measure preserving system $(X, \mathcal{B}, \mu, T)$ (or simply T) is said to be mildly mixing if there are no non-constant rigid functions in $L^2(X, \mathcal{B}, \mu)$.

For $\mathbf{Z}^k$-actions $\{T_{\mathbf{n}}\}_{\mathbf{n}\in\mathbf{Z}^k}$, one says that $f \in L^2(X, \mathcal{B}, \mu)$ is rigid if there exists a sequence $\{\mathbf{n}_l\}_{l=1}^\infty \subset \mathbf{Z}^k$ with $|\mathbf{n}_l| \to \infty$ (where $|\cdot|$ denotes the Euclidean norm) and $T_{\mathbf{n}_l} f \to f$ as $l \to \infty$. A system $(X, \mathcal{B}, \mu, \{T_{\mathbf{n}}\}_{\mathbf{n}\in\mathbf{Z}^k})$ is said to be mildly mixing if there are no non-constant rigid functions. In the present paper ramifications of mild mixing, or, properly speaking, a closely-related, more general form of mixing relative to a factor (see Definition 3.5), play a crucial role (as it does in [FK2] as well). Analogous to the fact that weak mixing implies weak mixing of higher orders, mild mixing implies mild mixing of higher orders. The following corollary of Theorem 4.10, which extends Theorem 4.8 of [B1], will be proved in Section 6. For its formulation, we need a notion of linear independence among IP-sets. A definition will be given later (Definition 6.7), but a sufficient condition (which is nearly general) for linear independence of the IP-sets $(n_\alpha^{(i)})_{\alpha\in\mathcal{F}}$, $1 \le i \le k$, is that for every $l_1, \cdots, l_k \in \mathbf{Z}$, there are at most finitely many $\alpha \in \mathcal{F}$ such that $l_1 n_\alpha^{(1)} + \cdots + l_k n_\alpha^{(k)} = 0$.

Theorem 0.14 Suppose that $(X, \mathcal{B}, \mu)$ is a probability space and $\{T_{\mathbf{n}}\}_{\mathbf{n}\in\mathbf{Z}^r}$ is a mildly mixing measure preserving $\mathbf{Z}^r$-action generated by $T_1, \cdots, T_r$. Suppose that $t \in \mathbf{N}$ and $p_{i,j}(x_1, \cdots, x_k) \in \mathbf{Q}[x_1, \cdots, x_k]$ with $p_{i,j}(\mathbf{Z}^k) \subset \mathbf{Z}$, $1 \le i \le r$, $1 \le j \le t$ are polynomials having the property that for any $1 \le j_1 \ne j_2 \le t$, the functions

$$(l_1, \cdots, l_k) \to \big(p_{1,j_1}(l_1, \cdots, l_k), \cdots, p_{r,j_1}(l_1, \cdots, l_k)\big)$$

and

$$\begin{aligned}&(l_1, \cdots, l_k)\\ \to &\big(p_{1,j_1}(l_1, \cdots, l_k) - p_{1,j_2}(l_1, \cdots, l_k), \cdots, p_{r,j_1}(l_1, \cdots, l_k) - p_{r,j_2}(l_1, \cdots, l_k)\big)\end{aligned}$$

are not constant. Suppose that $A_0, A_1, A_2, \cdots, A_t \in \mathcal{B}$ with $\mu(A_i) > 0$, $0 \le i \le t$. Then for any $\epsilon > 0$ and any linearly independent IP-sets $(n_\alpha^{(1)})_{\alpha \in \mathcal{F}}, \cdots, (n_\alpha^{(k)})_{\alpha \in \mathcal{F}}$ in $\mathbf{Z}$ there exists $\alpha \in \mathcal{F}$ such that

$$\left|\mu\Big(A_0 \cap \bigcap_{j=1}^{t} \Big(\prod_{i=1}^{r} T_i^{p_{i,j}(n_\alpha^{(1)}, \cdots, n_\alpha^{(k)})}\Big)^{-1} A_i\Big) - \prod_{i=0}^{r} \mu(A_i)\right| < \epsilon.$$

CHAPTER 1

FORMULATION OF MAIN THEOREM

Recall that we use $\mathcal{F}$ to denote the family of non-empty finite subsets of $\mathbf{N}$. We let $\mathcal{F}_\emptyset = \mathcal{F} \cup \{\emptyset\}$. One of the reasons that we are concerned with these objects is that in order to deal nicely with IP-sets, it is useful to see them as sequences indexed by either $\mathcal{F}$ or by $\mathcal{F}_\emptyset$ (this will be explained presently). Any sequence indexed by $\mathcal{F}$ (or by $\mathcal{F}_\emptyset$) will be called an $\mathcal{F}$-sequence. If G is an abelian group then any $\mathcal{F}$-sequence $(n_\alpha)_{\alpha\in\mathcal{F}} \subset G$ for which $n_{\alpha\cup\beta} = n_\alpha + n_\beta$ whenever $\alpha \cap \beta = \emptyset$ will be called an *IP-sequence.* We will now show that IP-sets may be indexed in a natural way by $\mathcal{F}$, whereupon they become IP-sequences. Recall that the IP-set generated by the sequence $G = \{g_i : i \in \mathbf{N}\} \subset \mathbf{Z}$ is given by

$$\Gamma = \{g_{i_1} + g_{i_2} + \cdots + g_{i_k} : i_1 < i_2 < \cdots < i_k, k \in \mathbf{N}\}.$$

If we set

$$n_{\{i_1,i_2,\cdots,i_k\}} = g_{i_1} + g_{i_2} + \cdots + g_{i_k}$$

then $(n_\alpha)_{\alpha\in\mathcal{F}}$ is an IP-sequence (that is, $n_{\alpha\cup\beta} = n_\alpha + n_\beta$ whenever $\alpha \cap \beta = \emptyset$) with the property that $\{n_\alpha : \alpha \in \mathcal{F}\} = \Gamma$. Henceforth, whenever we will be dealing with IP-sets we will assume that they have the structure of an IP-sequence. However, in a mild abuse of terminology, we may sometimes still refer to these IP-sequences as IP-sets. Furthermore, any time we have an IP-sequence $(n_\alpha)_{\alpha\in\mathcal{F}}$ given, we will, if needed, extend the indexing set to $\mathcal{F}_\emptyset$ by letting $n_\emptyset = 0$.

Suppose that $\alpha, \beta \in \mathcal{F}_\emptyset$ and suppose that for all $a \in \alpha$ and $b \in \beta$ we have $a < b$. In this case, we will write $\alpha < \beta$. (Notice that $\emptyset < \alpha < \emptyset$ for all $\alpha \in \mathcal{F}$.)

Of paramount importance to us is a notion of convergence for $\mathcal{F}$-sequences which will be defined later in this section. For this mode of convergence, any $\mathcal{F}$-sequence in a compact topological space will converge along a sub-sequence. In order to explain what is meant by a sub-sequence of an $\mathcal{F}$-sequence, we introduce, following [FK2], the notion of an *IP-ring.* Suppose that we are given a conventional sequence $(\alpha_i)_{i\in\mathbf{N}} \subset \mathcal{F}$ with $\alpha_1 < \alpha_2 < \alpha_3 < \cdots$. Let

$$\mathcal{F}^{(1)} = \Big\{ \bigcup_{i\in\beta} \alpha_i : \beta \in \mathcal{F} \Big\}.$$

Then $\mathcal{F}^{(1)}$ is called an IP-ring, as is $\mathcal{F}^{(1)}_\emptyset = \mathcal{F}^{(1)} \cup \{\emptyset\}$. Notice that the map

$$\xi : \mathcal{F} \to \mathcal{F}^{(1)}, \quad \xi(\alpha) = \bigcup_{i\in\alpha} \alpha_i$$

is bijective and structure preserving in the sense that $\xi(\alpha \cup \beta) = \xi(\alpha) \cup \xi(\beta)$. In particular, since $\mathcal{F}^{(1)}$ has the structure of $\mathcal{F}$, any sequence $(y_\alpha)_{\alpha\in\mathcal{F}^{(1)}}$ indexed by

$\mathcal{F}^{(1)}$ has the structure of an $\mathcal{F}$-sequence and indeed can be identified naturally with a particular $\mathcal{F}$-sequence, namely the $\mathcal{F}$-sequence $(x_\alpha)_{\alpha\in\mathcal{F}}$, where $x_\alpha = y_{\xi(\alpha)}$.

Finally, for $m \in \mathbf{N}$ we will denote by $\mathcal{F}^m_<$ $((\mathcal{F}_\emptyset)^m_<)$ the set of all m-tuples $(\alpha_1, \cdots, \alpha_m)$ in $\mathcal{F}^m$ $(\mathcal{F}^m_\emptyset)$ such that $\alpha_i < \alpha_j$ whenever $1 \le i < j \le m$.

Definition 1.1 Suppose that $k, r \in \mathbf{N}$ are fixed. Let $\mathbf{PE}(1)$ denote the set of expressions of the form

$$T(\alpha) = \prod_{i=1}^{r} T_i^{p_i(n_\alpha^{(1)}, \cdots, n_\alpha^{(k)})}, \quad \alpha \in \mathcal{F}_\emptyset, \tag{1.1}$$

where $p_i(x_1, \cdots, x_k) \in \mathbf{Z}[x_1, \cdots, x_k]$ with $p_i(0, 0, \cdots, 0) = 0$, $1 \le i \le r$. ($\mathbf{PE}$ stands for "polynomial expression".)

For $m \in \mathbf{N}$, let $\mathbf{PE}(m)$ denote the set of formal expressions having the form

$$T(\alpha_1, \cdots, \alpha_m) = \prod_{i=1}^{r} T_i^{p_i\left((n_{\alpha_j}^{(b)})_{1\le b\le k,\ 1\le j\le m}\right)}, \quad (\alpha_1, \cdots, \alpha_m) \in (\mathcal{F}_\emptyset)^m_<,$$

where p_i is a polynomial in a $k \times m$ matrix of variables having integer coefficients and zero constant term, $1 \le i \le r$. For $d \in \mathbf{N}$, we denote by $\mathbf{PE}_{\le d}(m)$ the subset of $\mathbf{PE}(m)$ consisting of such expressions constructed with polynomials p_i of degree $\le d$. (For our purposes, we take the degree of a monomial to be the sum of the powers of the variables appearing in it. Any polynomial can be uniquely written as a sum of monomials, no two of which are non-zero constant multiples of each other. The degree of a polynomial so expressed is the maximum degree of the corresponding monomials.)

The reader may notice that while the theorems appearing in the introduction concern themselves with polynomials $p(x) \in \mathbf{Q}[x]$ for which $p(\mathbf{Z}) \subset \mathbf{Z}$, the expressions of Definition 1.1, which are precisely the expressions dealt with in our main theorem, Theorem 1.3, are built with polynomials $p(x) \in \mathbf{Z}[x]$. The reason for this is that polynomials in $\mathbf{Z}[x]$ are easier to deal with, and the case of polynomials in $\mathbf{Q}[x]$ is easily reducible to the case of polynomials in $\mathbf{Z}[x]$. This reduction will be carried out in Section 6.

Example 1.2 Suppose for the moment that $k = 2$ and $r = 2$. Taking care not to confuse superscripts with exponents, one checks that the following is a member of $\mathbf{PE}(3)$:

$$T(\alpha_1, \alpha_2, \alpha_3) = T_1^{(n_{\alpha_1}^{(1)})^2 + n_{\alpha_1}^{(2)} n_{\alpha_2}^{(1)} - n_{\alpha_1}^{(2)} (n_{\alpha_3}^{(1)})^2 + 5 n_{\alpha_1}^{(1)} n_{\alpha_2}^{(1)} n_{\alpha_3}^{(2)}} T_2^{3(n_{\alpha_2}^{(2)})^3 - 2(n_{\alpha_1}^{(1)})^2 n_{\alpha_1}^{(2)} - 17(n_{\alpha_3}^{(1)})^2}.$$

We will now make a few observations regarding Definition 1.1. First of all, notice that the object $\mathbf{PE}(m)$ depends not only on m, but on k and r as well. The reason that k and r are suppressed in the notation is that they will be fixed natural numbers throughout the entire proof of Theorem 1.3. The notion of $\mathbf{PE}(m)$ depends only on k, r, and m, however. The symbols T_i, $1 \le i \le r$, and $n_\alpha^{(i)}$, $1 \le i \le k$, are merely placeholders. (We will now explain what this means.)

In essence, elements of $\mathbf{PE}(m)$ are just r-tuples of polynomials of km variables having integer coefficients and zero constant term. In particular, the element of $\mathbf{PE}(1)$ given by (1.1) corresponds to the r-tuple

$$\big(p_1(x_1, \cdots, x_k), \cdots, p_r(x_1, \cdots, x_k)\big).$$

The reason we write members of $\mathbf{PE}(m)$ the way we do is to indicate how they are used, namely to generate sequences of measure preserving transformations indexed by $\left(\mathcal{F}_\emptyset\right)_<^m$. (For example, suppose that $T_1, T_2, \cdots, T_r$ are commuting measure preserving transformations of a probability space $(X, \mathcal{B}, \mu)$ generating a group Ω, and suppose that $(n_\alpha^{(1)})_{\alpha\in\mathcal{F}}, \cdots, (n_\alpha^{(k)})_{\alpha\in\mathcal{F}}$ are IP-sets. Then (taking the case $m = 1$) any $T(\alpha) \in \mathbf{PE}(1)$ gives rise via equation (1.1) to an $\mathcal{F}$-sequence in Ω.) In a slight abuse of terminology, we will generally identify polynomial expressions with the $\mathcal{F}$-sequences of measure preserving transformations they give rise to.

However, we point out that $\mathbf{PE}(m)$ has a natural group structure which is independent of any specific family of transformations, or for that matter any fixed IP-sets. Given k, r, m, and $d \in \mathbf{N}$, $\mathbf{PE}_{\leq d}(m)$ is a finitely generated, free abelian group under the multiplicative operation which corresponds to addition on the underlying r-tuples of polynomials. (If the members of $\mathbf{PE}(m)$ are taken to be transformations, then this product is of course just composition.) In $\mathbf{PE}_{\leq d}(1)$, for example, this multiplication may be written

$$\left(\prod_{i=1}^r T_i^{p_i(n_\alpha^{(1)},\cdots,n_\alpha^{(k)})}\right)\left(\prod_{i=1}^r T_i^{q_i(n_\alpha^{(1)},\cdots,n_\alpha^{(k)})}\right) = \left(\prod_{i=1}^r T_i^{p_i(n_\alpha^{(1)},\cdots,n_\alpha^{(k)})+q_i(n_\alpha^{(1)},\cdots,n_\alpha^{(k)})}\right).$$

An example of this multiplication in $\mathbf{PE}(2)$ is

$$\begin{aligned}&\left(T_1^{(n_{\alpha_1}^{(1)})^2+(n_{\alpha_2}^{(2)})^4} T_2^{6(n_{\alpha_1}^{(2)})^1 7(n_{\alpha_2}^{(1)})^4}\right)\left(T_1^{-2(n_{\alpha_1}^{(2)})^3(n_{\alpha_1}^{(1)})^2} T_2^{-(n_{\alpha_1}^{(2)})^1 7(n_{\alpha_2}^{(1)})^4}\right)\\ =&T_1^{(n_{\alpha_1}^{(1)})^2+(n_{\alpha_2}^{(2)})^4-2(n_{\alpha_1}^{(2)})^3(n_{\alpha_1}^{(1)})^2} T_2^{5(n_{\alpha_1}^{(2)})^1 7(n_{\alpha_2}^{(1)})^4}.\end{aligned}$$

The reason we don't fix transformations or IP-sets in Definition 1.1 is that we want to avoid non-trivial indentities, or "torsion", in $\mathbf{PE}(m)$. For some choices of probability space and transformations T_i, two different polynomial expressions in $\mathbf{PE}(1)$ may give rise to the same $\mathcal{F}$-sequence in Ω owing to non-trivial identities occuring among the transformations. For example, if $T_1 = T_2$, then the polynomial expression $T(\alpha) = T_1^{n_\alpha^{(1)}} T_2^{-n_\alpha^{(1)}}$ yields a function which sends every $\alpha \in \mathcal{F}$ to the identity transformation $I \in \Omega$, yet $T(\alpha)$ is *not* the identity element in $\mathbf{PE}(1)$. Similarly, having fixed IP-sets $(n_\alpha^{(i)})_{\alpha\in\mathcal{F}}$, $1 \leq i \leq k$, one may find that there exist non-trivial identities among them, but these do not give rise to identities in $\mathbf{PE}(m)$. For example, if $n_\alpha^{(1)} + n_\alpha^{(2)} = n_\alpha^{(3)}$ for all $\alpha \in \mathcal{F}$ then $S(\alpha) = T_1^{n_\alpha^{(1)}+n_\alpha^{(2)}}$ and $U(\alpha) = T_1^{n_\alpha^{(3)}}$ are still distinct elements of $\mathbf{PE}(1)$ in spite of the fact that for *any* measure preserving system $(X, \mathcal{B}, \mu)$ and group of measure preserving transformations Ω generated by commuting T_i, $1 \leq i \leq r$, $S(\alpha)$ and $U(\alpha)$ induce the same $\mathcal{F}$-sequence in Ω. To conclude, $\mathbf{PE}_{\leq d}(m)$ is a finitely generated free abelian group depending only on k, r, m and d. If $k = r = 2$, then a basis for $\mathbf{PE}_{\leq 2}(1)$ is

$$\{T_1^{n_\alpha}, T_1^{n_\alpha^2}, T_1^{m_\alpha}, T_1^{m_\alpha^2}, T_1^{m_\alpha n_\alpha}, T_2^{n_\alpha}, T_2^{n_\alpha^2}, T_2^{m_\alpha}, T_2^{m_\alpha^2}, T_2^{m_\alpha n_\alpha}\}.$$

We are almost in position to formulate our main theorem. First, however, we need to introduce the aforementioned notion of *IP-convergence*. Suppose that (x_α) is an $\mathcal{F}$-sequence in a topological space and $\mathcal{F}^{(1)}$ is an IP-ring. We will write (see [FK2])

$$\operatorname*{IP-lim}_{\alpha\in\mathcal{F}^{(1)}} x_\alpha = z$$

if for every neighborhood W of z there exists $\beta \in \mathcal{F}$ having the property that for every $\alpha \in \mathcal{F}^{(1)}$ with $\alpha > \beta$ we have $x_\alpha \in W$.

The notion of IP-convergence has a natural multi-parameter generalization. Suppose that X is a topological space, $\mathcal{F}^{(1)}$ is an IP-ring, $m \in \mathbf{N}$, and

$$\{x_{(\alpha_1,\cdots,\alpha_m)} : (\alpha_1,\cdots,\alpha_m) \in (\mathcal{F}^{(1)})^m_<\}$$

is a sequence in X indexed by $(\mathcal{F}^{(1)})^m_<$. We shall write

$$\underset{(\alpha_1,\cdots,\alpha_m)\in(\mathcal{F}^{(1)})^m_<}{\text{IP-lim}} x_{(\alpha_1,\cdots,\alpha_m)} = z$$

if for any neighborhood U of z, there exists $\alpha_0 \in \mathcal{F}^{(1)}$ such that for all $(\alpha_1,\cdots,\alpha_m) \in (\mathcal{F}^{(1)})^m_<$, $\alpha_1 > \alpha_0$, we have $x_{(\alpha_1,\cdots,\alpha_m)} \in U$.

Here is our main theorem.

Theorem 1.3 Let $d, k, r \in \mathbf{N}$ and fix IP-sets $(n_\alpha^{(i)})_{\alpha\in\mathcal{F}}$, $1 \le i \le k$. Let $\mathbf{PE}_{\le d}(m)$ be as in Definition 1.1 and suppose that $(X, \mathcal{A}, \mu, \Omega)$ is a measure preserving system, where Ω is generated by commuting, invertible transformations $T_1, \cdots, T_r$. For every $A \in \mathcal{A}$ with $\mu(A) > 0$ and every $m, t \in \mathbf{N}$ there exist an IP-ring $\mathcal{F}^{(1)}$ and a number $a = a(A, m, t, d) > 0$ having the property that for every set of polynomial expressions $\{S_0, \cdots, S_t\} \subset \mathbf{PE}_{\le d}(m)$ we have

$$\underset{(\alpha_1,\cdots,\alpha_m)\in(\mathcal{F}^{(1)})^m_<}{\text{IP-lim}} \mu\Big(\bigcap_{i=0}^{t} S_i(\alpha_1,\cdots,\alpha_m)^{-1}A\Big) \ge a.$$

Sections 2-5 of this paper constitute the proof of Theorem 1.3. We note that k and r will be fixed at the beginning of Section 2 and the IP-sets $(n_\alpha)_{\alpha\in\mathcal{F}}$ will be fixed before Example 2.2. The system $(X, \mathcal{A}, \mu, \Omega)$ will not be fixed until the beginning of Section 5.

The strategy of the proof we will employ could be described as a polynomialization of the proof of the linear IP-Szemerédi theorem in [FK2]. The general idea (also central to the proofs of the increasingly stronger results in [F], [FK1] and [BL1]) is to exhaust the measure preserving system by a chain of special factors in which the two complementary notions of compactness and weak mixing are controllably combined.

Specifically, one shows that if the main theorem holds for a factor $(Y, \mathcal{B}, \nu, \Omega)$ of our measure preserving system $(X, \mathcal{A}, \mu, \Omega)$ then it also holds for some non-trivial extension of $(Y, \mathcal{B}, \nu, \Omega)$ (which is again a factor of $(X, \mathcal{A}, \mu, \Omega)$), and that furthermore the set of factors for which the main theorem holds has a maximal element which therefore coincides with $(X, \mathcal{A}, \mu, \Omega)$.

The structure of the paper is as follows: in Section 2 we establish some terminology and notation and introduce the main combinatorial tools to be use in subsequent sections, namely Hindman's theorem ([H]) and the polynomial Hales-Jewett theorem ([BL2]). Section 3 is devoted to definitions and main properties of what we call, in keeping with the terminological tradition of [FK1] and [FK2], *primitive extensions*. These extensions are exactly the ones we will be dealing with in extending the validity of the main theorem from a factor to a non-trivial extension. In Section 4 we deal with the phenomenon of *relative mixing* and show inductively that this fundamental property of extensions ensures relative polynomial mixing of all orders. The proof is brought to completion in Section 5, in which it is shown how the two phenomena of compactness and mixing combine to push the validity of our main theorem from a measure preserving system to any primitive extension. Section 6 is devoted to applications of the main theorem to ergodic theory, more specifically to a fine analysis of the structure of the set of multiple recurrence return times of weakly and mildly mixing measure preserving systems. Section 7 concentrates on combinatorial applications, and Section 8 is a short reflection on the possibilities for obtaining stronger results than those we obtain here. Finally, in an appendix we prove a multi-parameter weakly mixing ergodic theorem which is needed for one of the applications of Section 6.

CHAPTER 2

PRELIMINARIES

This preparatory chapter has three sections. The first one attempts to introduce a well-organized notation which will help us later to deal without unnecessary confusion with the groups $\mathbf{PE}_{\leq d}(m)$ of polynomial expressions defined in Section 1. The second one introduces the combinatorial machinery we will be using in the proof of Theorem 1.3, namely the polynomial Hales-Jewett theorem ([BL2]). The third and final part is devoted to a very brief account of some generalities concerning factors and the disintegration of measures.

This section, together with the next three sections, constitute a proof of Theorem 1.3. We now select and fix for the whole of this proof $k \in \mathbf{N}$ and $r \in \mathbf{N}$. It is these fixed values of k and r which will be assumed in Definition 1.1, the definition of $\mathbf{PE}(m)$.

2.1. Notation, Definitions and Examples.

Note that if $T \in \mathbf{PE}(m)$, then T is uniquely expressible in the form

$$T(\alpha_1, \cdots, \alpha_m) = S(\alpha_1)S^{(\alpha_1)}(\alpha_2) \cdots S^{(\alpha_1, \cdots, \alpha_{m-1})}(\alpha_m), \tag{2.1}$$

where $S(\alpha_1)$ is the portion of $T(\alpha_1, \cdots, \alpha_m)$ which depends solely on α_1, $S^{(\alpha_1)}(\alpha_2)$ is the portion of $T(\alpha_1, \cdots, \alpha_m)$ which depends only on α_1 and α_2, but not solely on α_1, and so on. To see exactly what this means, recall that T has the form

$$T(\alpha_1, \cdots, \alpha_m) = \prod_{i=1}^{r} T_i^{p_i\left((n_{\alpha_j}^{(b)})_{1 \leq b \leq k,\ 1 \leq j \leq m}\right)}, \quad (\alpha_1, \cdots, \alpha_m) \in (\mathcal{F}_\emptyset)_<^m,$$

where for all i, $1 \leq i \leq r$, p_i is a polynomial in the km "variables" $n_{\alpha_j}^{(b)}$, $1 \leq b \leq k$, $1 \leq j \leq m$. Equation (2.1) corresponds to the separation of these polynomials into monomials and recombining these monomials according to the highest value of j for which α_j is appearing in them. For example, we get $S(\alpha_1)$ by keeping only the monomials in the "variables" $n_{\alpha_1}^{(b)}$, $1 \leq b \leq k$. Alternatively,

$$S(\alpha_1) = T(\alpha_1, \emptyset, \cdots, \emptyset).$$

$S^{(\alpha_1)}(\alpha_2)$ comes from the monomials in $n_{\alpha_j}^{(b)}$, $1 \leq j \leq 2$, $1 \leq b \leq k$ which contain at least one occurence of some "variable" $n_{\alpha_2}^{(b)}$. Or,

$$S^{(\alpha_1)}(\alpha_2) = T(\alpha_1, \alpha_2, \emptyset, \cdots, \emptyset)S(\alpha_1)^{-1}.$$

Thus $S^{(\alpha_1,\alpha_2)}(\alpha_3) = T(\alpha_1, \alpha_2, \alpha_3, \emptyset, \cdots, \emptyset)\left(S^{(\alpha_1)}(\alpha_2)S(\alpha_1)\right)^{-1}$, and so on.

Example 2.1 Suppose again that

$$T(\alpha_1, \alpha_2, \alpha_3) = T_1^{(n_{\alpha_1}^{(1)})^2 + n_{\alpha_1}^{(2)} n_{\alpha_2}^{(1)} - n_{\alpha_1}^{(2)} (n_{\alpha_3}^{(1)})^2 + 5 n_{\alpha_1}^{(1)} n_{\alpha_2}^{(1)} n_{\alpha_3}^{(2)}} T_2^{3(n_{\alpha_2}^{(2)})^3 - 2(n_{\alpha_1}^{(1)})^2 n_{\alpha_1}^{(2)} - 17(n_{\alpha_3}^{(1)})^2}.$$

Then the expressions S, $S^{(\alpha_1)}$, and $S^{(\alpha_1,\alpha_2)}$ of equation (2.1) are given by

$$\begin{aligned} S(\alpha_1) &= T_1^{(n_{\alpha_1}^{(1)})^2} T_2^{-2(n_{\alpha_1}^{(1)})^2 n_{\alpha_1}^{(2)}} \\ S^{(\alpha_1)}(\alpha_2) &= T_1^{n_{\alpha_1}^{(2)} n_{\alpha_2}^{(1)}} T_2^{3(n_{\alpha_2}^{(2)})^3} \\ S^{(\alpha_1,\alpha_2)}(\alpha_3) &= T_1^{n_{\alpha_1}^{(2)} (n_{\alpha_3}^{(1)})^2 + 5 n_{\alpha_1}^{(1)} n_{\alpha_2}^{(1)} n_{\alpha_3}^{(2)}} T_2^{-17(n_{\alpha_3}^{(1)})^2}. \end{aligned} \tag{2.2}$$

We now fix the IP-sets $(n_\alpha^{(i)})_{\alpha \in \mathcal{F}}$, $1 \leq i \leq k$ which occur in the statement of Theorem 1.3. Having done this, note that for any fixed $(\alpha_1, \cdots, \alpha_{m-1}) \in (\mathcal{F}_\emptyset)^{m-1}$ we have, plugging in actual numbers for the "variables" (which we can stop thinking of as variables now that these are fixed IP-sets) $n_{\alpha_j}^{(b)}$,

$$\{S(\cdot), S^{(\alpha_1)}(\cdot), \cdots, S^{(\alpha_1, \cdots, \alpha_{m-1})}(\cdot)\} \subset \mathbf{PE}(1). \tag{2.3}$$

Example 2.2 Suppose that $\alpha_1, \alpha_2 \in \mathcal{F}$ have been fixed and suppose that (for instance) $n_{\alpha_1}^{(1)} = 2$, $n_{\alpha_2}^{(1)} = 3$, and $n_{\alpha_1}^{(2)} = 4$. Then from the last line of (2.2) we get

$$S^{(\alpha_1,\alpha_2)}(\alpha_3) = T_1^{4(n_{\alpha_3}^{(1)})^2 + 30 n_{\alpha_3}^{(2)}} T_2^{-17(n_{\alpha_3}^{(1)})^2}.$$

In the course of our proof of Theorem 1.3 we will often be dealing with a fixed subgroup G of $\mathbf{PE}_{\leq d}(1)$ having the property that any $T \in G$ exhibits some form of (relative) compactness (by this we mean that the family of operators $\{T(\alpha) : \alpha \in \mathcal{F}^{(1)}\}$ exhibits compactness along fibers–see Section 3). The following definition arises out of the need to isolate those members

$$T(\alpha_1, \cdots, \alpha_m) = S(\alpha_1) S^{(\alpha_1)}(\alpha_2) \cdots S^{(\alpha_1, \cdots, \alpha_{m-1})}(\alpha_m)$$

of $\mathbf{PE}(m)$ which exhibit a similar kind of relative compactness by virtue of the fact that along some IP-ring the set appearing in (2.3) is eventually contained in G whenever $\alpha_1 < \alpha_2 < \cdots < \alpha_{m-1}$.

Definition 2.3 Suppose $d \in \mathbf{N}$, G is a subgroup of $\mathbf{PE}_{\leq d}(1)$, and $\mathcal{F}^{(1)}$ is an IP-ring. Suppose that $T \in \mathbf{PE}(m)$ has decomposition

$$T(\alpha_1, \cdots, \alpha_m) = S(\alpha_1) S^{(\alpha_1)}(\alpha_2) \cdots S^{(\alpha_1, \cdots, \alpha_{m-1})}(\alpha_m),$$

as in equation (2.1). We will write $T \in \mathbf{PE}(G, \mathcal{F}^{(1)})$ if $S \in G$, and if there exists $\alpha_0 \in \mathcal{F}^{(1)}$ having the property that whenever $(\alpha_1, \cdots, \alpha_{m-1}) \in (\mathcal{F}^{(1)})_<^{m-1}$ with $\alpha_0 < \alpha_1$ we have

$$\{S^{(\alpha_1)}, \cdots, S^{(\alpha_1, \cdots, \alpha_{m-1})}\} \subset G.$$

In the sequel, we will make use of the fact that $\mathbf{PE}(G, \mathcal{F}^{(1)})$ is a group, as well as of the following result, whose proof we omit:

Proposition 2.4 Suppose that $m < N$, $T(\alpha_1, \cdots, \alpha_m) \in \mathbf{PE}(G, \mathcal{F}^{(1)})$ and $1 \leq n_1 < \cdots < n_m \leq N$. Let $S \in \mathbf{PE}(N)$ be defined by $S(\alpha_1, \cdots, \alpha_N) = T(\alpha_{n_1}, \cdots, \alpha_{n_m})$. Then $S \in \mathbf{PE}(G, \mathcal{F}^{(1)})$ as well.

Suppose that $T \in \mathbf{PE}(1)$ is given by

$$T(\alpha) = \prod_{i=1}^{r} T_i^{p_i(n_\alpha^{(1)}, \cdots, n_\alpha^{(k)})}, \quad \alpha \in \mathcal{F}_\emptyset.$$

Let

$$S(\alpha_1, \cdots, \alpha_m) = \prod_{i=1}^{r} T_i^{p_i\left(n_{\alpha_1}^{(1)} + \cdots + n_{\alpha_m}^{(1)}, \cdots, n_{\alpha_1}^{(k)} + \cdots + n_{\alpha_m}^{(k)}\right)}, \quad (\alpha_1, \cdots, \alpha_m) \in (\mathcal{F}_\emptyset)^m_<. \tag{2.5}$$

Then $S(\alpha_1, \cdots, \alpha_m) \in \mathbf{PE}(m)$. More importantly, $S(\alpha_1, \cdots, \alpha_m) = T(\alpha_1 \cup \cdots \cup \alpha_m)$ whenever $(\alpha_1, \cdots, \alpha_m) \in (\mathcal{F}_\emptyset)^m_<$. Therefore it makes sense for us to write $T(\alpha_1 \cup \cdots \cup \alpha_m) \in \mathbf{PE}(m)$.

In order to analyze the expression $T(\alpha_1 \cup \cdots \cup \alpha_m)$, we will need to break it down in a manner analogous to (but even finer than) equation (2.1).

Definition 2.5 If $T \in \mathbf{PE}(1)$, write $T^{(0)} = I$, $T^{(1)} = T$, and let $T^{(2)} \in \mathbf{PE}(2)$ be determined by $T(\alpha_1 \cup \alpha_2) = T(\alpha_1)T(\alpha_2)T^{(2)}(\alpha_1, \alpha_2)$, $(\alpha_1, \alpha_2) \in (\mathcal{F}_\emptyset)^2_<$. Inductively define $T^{(k)}$ for $k \geq 3$ by

$$T(\alpha_1 \cup \cdots \cup \alpha_k) = \prod_{\{\beta_1, \cdots, \beta_j\} \subset \{\alpha_1, \cdots, \alpha_k\}} T^{(j)}(\beta_1, \cdots, \beta_j), \quad (\alpha_1, \cdots, \alpha_m) \in (\mathcal{F}_\emptyset)^m_<. \tag{2.6}$$

Example 2.6 If $k = 3$ then (2.6) becomes

$$\begin{aligned} &T(\alpha_1 \cup \alpha_2 \cup \alpha_3) \\ =&T(\alpha_1)T(\alpha_2)T(\alpha_3)T^{(2)}(\alpha_1, \alpha_2)T^{(2)}(\alpha_1, \alpha_3)T^{(2)}(\alpha_2, \alpha_3)T^{(3)}(\alpha_1, \alpha_2, \alpha_3). \end{aligned} \tag{2.7}$$

Taking $T(\alpha) = T_1^{2(n_\alpha^{(1)})^2 n_\alpha^{(2)}} T_2^{n_\alpha^{(1)} - 5(n_\alpha^{(2)})^2}$ in (2.7) gives us

$$\begin{aligned}
&T(\alpha_1 \cup \alpha_2 \cup \alpha_3) \\
=&\left(T_1^{2(n_{\alpha_1}^{(1)} + n_{\alpha_2}^{(1)} + n_{\alpha_3}^{(1)})^2 (n_{\alpha_1}^{(2)} + n_{\alpha_2}^{(2)} + n_{\alpha_3}^{(2)})} T_2^{n_{\alpha_1}^{(1)} + n_{\alpha_2}^{(1)} + n_{\alpha_3}^{(1)} - 5(n_{\alpha_1}^{(2)} + n_{\alpha_2}^{(2)} + n_{\alpha_3}^{(2)})^2}\right) \\
=&\left(T_1^{2(n_{\alpha_1}^{(1)})^2 n_{\alpha_1}^{(2)}} T_2^{n_{\alpha_1}^{(1)} - 5(n_{\alpha_1}^{(2)})^2}\right) \\
&\left(T_1^{2(n_{\alpha_2}^{(1)})^2 n_{\alpha_2}^{(2)}} T_2^{n_{\alpha_2}^{(1)} - 5(n_{\alpha_2}^{(2)})^2}\right) \\
&\left(T_1^{2(n_{\alpha_3}^{(1)})^2 n_{\alpha_3}^{(2)}} T_2^{n_{\alpha_3}^{(1)} - 5(n_{\alpha_3}^{(2)})^2}\right) \\
&\left(T_1^{2(n_{\alpha_1}^{(1)})^2 n_{\alpha_2}^{(2)} + 4n_{\alpha_1}^{(1)} n_{\alpha_2}^{(1)} n_{\alpha_2}^{(2)} + 2(n_{\alpha_2}^{(1)})^2 n_{\alpha_1}^{(2)} + 4n_{\alpha_1}^{(1)} n_{\alpha_2}^{(1)} n_{\alpha_1}^{(2)}} T_2^{-10 n_{\alpha_1}^{(2)} n_{\alpha_2}^{(2)}}\right) \\
&\left(T_1^{2(n_{\alpha_1}^{(1)})^2 n_{\alpha_3}^{(2)} + 4n_{\alpha_1}^{(1)} n_{\alpha_3}^{(1)} n_{\alpha_3}^{(2)} + 2(n_{\alpha_3}^{(1)})^2 n_{\alpha_1}^{(2)} + 4n_{\alpha_1}^{(1)} n_{\alpha_3}^{(1)} n_{\alpha_1}^{(2)}} T_2^{-10 n_{\alpha_1}^{(2)} n_{\alpha_3}^{(2)}}\right) \\
&\left(T_1^{2(n_{\alpha_3}^{(1)})^2 n_{\alpha_2}^{(2)} + 4n_{\alpha_3}^{(1)} n_{\alpha_2}^{(1)} n_{\alpha_2}^{(2)} + 2(n_{\alpha_2}^{(1)})^2 n_{\alpha_3}^{(2)} + 4n_{\alpha_3}^{(1)} n_{\alpha_2}^{(1)} n_{\alpha_3}^{(2)}} T_2^{-10 n_{\alpha_3}^{(2)} n_{\alpha_2}^{(2)}}\right) \\
&\left(T_1^{4(n_{\alpha_1}^{(1)} n_{\alpha_2}^{(1)} n_{\alpha_3}^{(2)} + n_{\alpha_1}^{(1)} n_{\alpha_3}^{(1)} n_{\alpha_2}^{(2)} + n_{\alpha_2}^{(1)} n_{\alpha_3}^{(1)} n_{\alpha_1}^{(2)})}\right).
\end{aligned}$$

We remark that if $T \in \mathbf{PE}_{\leq d}(1)$ then $T^{(k)} = I$ for $k > d$.

Suppose now that $d \in \mathbf{N}$ and $T \in \mathbf{PE}_{\leq d}(m)$, where $m > 1$. We will define new polynomial expressions

$$T^{(a_1,\cdots,a_m)} \in \mathbf{PE}(a_1 + \cdots + a_m)$$

for any m-tuple of non-negative integers $(a_1, \cdots, a_m)$ such that $T^{(a_1,\cdots,a_m)} = I$ whenever

$$a_1 + \cdots + a_m > d.$$

We begin with a few starting cases. First set $T^{(0,0,\cdots,0)} = I$. Let $T^{(1,0,0,\cdots,0)}(\alpha_1)$ be the part of $T(\alpha_1, \cdots, \alpha_m)$ which depends only on α_1. That is,

$$T^{(1,0,\cdots,0)}(\alpha_1) = T(\alpha_1, \emptyset, \emptyset, \cdots, \emptyset).$$

Let $T^{(0,1,0,\cdots,0)}$ be the part of $T(\alpha_1, \cdots, \alpha_m)$ which depends only on α_2,

$$T^{(0,1,0,\cdots,0)}(\alpha_2) = T(\emptyset, \alpha_2, \emptyset, \cdots, \emptyset),$$

and so on. Let $T^{(1,1,0,\cdots,0)}(\alpha_1, \alpha_2)$ be the part of $T(\alpha_1, \cdots, \alpha_m)$ depending only on α_1 and α_2, but on neither alone, that is, let

$$T^{(1,1,0,\cdots,0)}(\alpha_1, \alpha_2) = T(\alpha_1, \alpha_2, \emptyset, \cdots, \emptyset)\big(T^{(1,0,\cdots,0)}(\alpha_1)T^{(0,1,0,\cdots,0)}(\alpha_2)\big)^{-1}.$$

Other expressions $T^{(a_1,\cdots,a_m)}$, $a_i \in \{0,1\}$, are defined analogously.

Example 2.7 Suppose again that

$$T(\alpha_1, \alpha_2, \alpha_3) = T_1^{(n^{(1)}_{\alpha_1})^2 + n^{(2)}_{\alpha_1} n^{(1)}_{\alpha_2} - n^{(2)}_{\alpha_1}(n^{(1)}_{\alpha_3})^2 + 5n^{(1)}_{\alpha_1} n^{(1)}_{\alpha_2} n^{(2)}_{\alpha_3}} T_2^{3(n^{(2)}_{\alpha_2})^3 - 2(n^{(1)}_{\alpha_1})^2 n^{(2)}_{\alpha_1} - 17(n^{(1)}_{\alpha_3})^2}.$$

Then

$$\begin{aligned}
T^{(1,0,0)}(\alpha_1) &= T_1^{(n^{(1)}_{\alpha_1})^2} T_2^{-2(n^{(1)}_{\alpha_1})^2 n^{(2)}_{\alpha_1}} \\
T^{(0,1,0)}(\alpha_2) &= T_2^{3(n^{(2)}_{\alpha_2})^3} \\
T^{(1,1,0)}(\alpha_1, \alpha_2) &= T_1^{n^{(2)}_{\alpha_1} n^{(1)}_{\alpha_2}} \\
T^{(0,0,1)}(\alpha_3) &= T_2^{-17(n^{(1)}_{\alpha_3})^2} \\
T^{(1,0,1)}(\alpha_1, \alpha_3) &= T_1^{n^{(2)}_{\alpha_1}(n^{(1)}_{\alpha_3})^2} \\
T^{(0,1,1)}(\alpha_2, \alpha_3) &= I \\
T^{(1,1,1)}(\alpha_1, \alpha_2, \alpha_3) &= T_1^{5n^{(1)}_{\alpha_1} n^{(1)}_{\alpha_2} n^{(2)}_{\alpha_3}}.
\end{aligned}$$

Now, $T^{(2,0,\cdots,0)}(\alpha_1^{(1)}, \alpha_1^{(2)})$ is defined to be the part of $T(\alpha_1^{(1)} \cup \alpha_1^{(2)}, \alpha_2, \cdots, \alpha_m)$ which depends on $\alpha_1^{(1)}$ and $\alpha_1^{(2)}$, but on neither alone. That is,

$$T^{(2,0,\cdots,0)}(\alpha_1^{(1)}, \alpha_1^{(2)}) = T(\alpha_1^{(1)} \cup \alpha_1^{(2)}, \emptyset, \cdots, \emptyset)\big(T^{(1,0,\cdots,0)}(\alpha_1^{(1)})T^{(1,0,\cdots,0)}(\alpha_1^{(2)})\big)^{-1}.$$

Example 2.8 For $T(\alpha_1, \alpha_2, \alpha_3)$ as above,

$$\begin{aligned}
&T^{(2,0,0)}(\alpha_1^{(1)}, \alpha_1^{(2)}) \\
=&T_1^{2n^{(1)}_{\alpha_1^{(1)}} n^{(1)}_{\alpha_1^{(2)}}} T_2^{2((n_{\alpha_1^{(1)}})^2 n^{(2)}_{\alpha_1^{(2)}} + (n^{(1)}_{\alpha_1^{(2)}})^2 n^{(2)}_{\alpha_1^{(1)}} + 2n^{(1)}_{\alpha_1^{(1)}} n^{(1)}_{\alpha_1^{(2)}} (n^{(2)}_{\alpha_1^{(1)}} + n^{(2)}_{\alpha_1^{(2)}}))}.
\end{aligned}$$

$T^{(2,1,0,\cdots,0)}(\alpha_1^{(1)},\alpha_1^{(2)},\alpha_2)$ is defined to be that part of

$$T(\alpha_1^{(1)}\cup\alpha_1^{(2)},\alpha_2,\alpha_3,\cdots,\alpha_m)$$

which depends on $\{\alpha_1^{(1)},\alpha_1^{(2)},\alpha_2\}$, but not on any proper subset of this set. In other words,

$$\begin{aligned}&T^{(2,1,0,\cdots,0)}(\alpha_1^{(1)},\alpha_1^{(2)},\alpha_2)\\ =&T(\alpha_1^{(1)}\cup\alpha_1^{(2)},\alpha_2,\emptyset,\cdots,\emptyset)\Big(T^{(2,0,\cdots,0)}(\alpha_1^{(1)},\alpha_1^{(2)})T^{(1,0,\cdots,0)}(\alpha_1^{(1)})\\ &\quad T^{(1,0,\cdots,0)}(\alpha_1^{(2)})T^{(0,1,0,\cdots,0)}(\alpha_2)T^{(1,1,0,\cdots,0)}(\alpha_1^{(1)},\alpha_2)T^{(1,1,0,\cdots,0)}(\alpha_1^{(2)},\alpha_2)\Big)^{-1}.\end{aligned}$$

Hopefully, the following definition will now be marginally accessible.

Definition 2.9 Suppose that $T(\alpha_1,\cdots,\alpha_m)\in\mathbf{PE}(m)$. Let $T^{(0,0,\cdots,0)}=I$, and for every m-tuple $(a_1,\cdots,a_m)$ of non-negative integers, at least one of which is not zero, let

$$T^{(a_1,\cdots,a_m)}(\alpha_1^{(1)},\alpha_1^{(2)},\cdots,\alpha_1^{(a_1)},\alpha_2^{(1)},\alpha_2^{(2)},\cdots,\alpha_2^{(a_2)},\cdots,\alpha_m^{(1)},\alpha_m^{(2)},\cdots,\alpha_m^{(a_m)})$$

be the portion of

$$T(\alpha_1^{(1)}\cup\alpha_1^{(2)}\cup\cdots\cup\alpha_1^{(a_1)},\alpha_2^{(1)}\cup\alpha_2^{(2)}\cup\cdots\cup\alpha_2^{(a_2)},\cdots,\alpha_m^{(1)}\cup\alpha_m^{(2)}\cup\cdots\cup\alpha_m^{(a_m)})$$

which depends on no proper subset of

$$\{\alpha_1^{(1)},\alpha_1^{(2)},\cdots,\alpha_1^{(a_1)},\alpha_2^{(1)},\alpha_2^{(2)},\cdots,\alpha_2^{(a_2)},\cdots,\alpha_m^{(1)},\alpha_m^{(2)},\cdots,\alpha_m^{(a_m)}\}.$$

In particular, Definition 2.9 tells us that

$$\begin{aligned}&T\big(\alpha_1^{(1)}\cup\alpha_1^{(2)}\cup\cdots\cup\alpha_1^{(a_1)},\alpha_2^{(1)}\cup\alpha_2^{(2)}\cup\cdots\cup\alpha_2^{(a_2)},\cdots\\ &\qquad\cdots,\alpha_m^{(1)}\cup\alpha_m^{(2)}\cup\cdots\cup\alpha_m^{(a_m)}\big)\\ =&\prod_{\substack{\{\beta_i^{(1)},\cdots,\beta_i^{(b_i)}\}\\ \subset\{\alpha_i^{(1)},\cdots,\alpha_i^{(a_i)}\},\\ 1\le i\le m}} T^{(b_1,\cdots,b_m)}\big(\beta_1^{(1)},\beta_1^{(2)},\cdots,\beta_1^{(b_1)},\beta_2^{(1)},\beta_2^{(2)},\cdots,\beta_2^{(b_2)},\cdots\\ &\qquad\cdots,\beta_m^{(1)},\beta_m^{(2)},\cdots,\beta_m^{(b_m)}\big).\end{aligned}\tag{2.8}$$

2.2 Combinatorial Tools.

We now introduce the combinatorial machinery we will need. The key tool for this is the following polynomial Hales-Jewett theorem ([BL2]).

Theorem 2.10 Suppose numbers $k,d,r\in\mathbf{N}$ are given. Then there exists a number $N=N(k,d,r)\in\mathbf{N}$ having the property that whenever we have an r-cell partition

$$F\Big(\{1,\cdots,k\}\times\{1,\cdots,N\}^d\Big)=\bigcup_{i=1}^r C_i,$$

one of the sets C_i, $1\le i\le r$ contains a configuration of the form

$$\Big\{A\cup(B\times S^d):B\subset\{1,\cdots,k\}\Big\}$$

for some $A \subset (\{1,\cdots,k\} \times \{1,\cdots,N\}^d)$ and some non-empty set $S \subset \{1,\cdots,N\}$ satisfying

$$A \cap (\{1,\cdots,k\} \times S^d) = \emptyset.$$

Definition 2.11 Given a subgroup $G < \mathbf{PE}_{\leq}(1)$, and an IP-ring $\mathcal{F}^{(1)}$, the pair $(G, \mathcal{F}^{(1)})$ is said to be *balanced* if for every $T(\alpha_1,\cdots,\alpha_m) \in \mathbf{PE}(G,\mathcal{F}^{(1)})$ (see Definition 2.3) and every $a_1,\cdots,a_m \in \mathbf{N} \cup \{0\}$, we have $T^{(a_1,\cdots,a_m)} \in \mathbf{PE}(G,\mathcal{F}^{(1)})$.

Here is the combinatorial result we need, which is a consequence of Theorem 2.10.

Theorem 2.12 Suppose $t,d,l,m \in \mathbf{N}$. There exist numbers $N = N(t,d,l)$ and $w(t,d,l) \in \mathbf{N}$ such that for any IP-ring $\mathcal{F}^{(1)}$ and subgroup $G < \mathbf{PE}_{\leq d}(1)$ with $(G,\mathcal{F}^{(1)})$ balanced, and any sets of polynomial expressions

$$\{R_i(\alpha_1,\cdots,\alpha_m)\}_{i=1}^t \subset \mathbf{PE}_{\leq d}(G,\mathcal{F}^{(1)})$$

and

$$\{W_i(\alpha_1,\cdots,\alpha_m)\}_{i=1}^t \subset \mathbf{PE}_{\leq d}(m),$$

there exist sets of polynomial expressions

$$L = \{L_i(\alpha_1,\cdots,\alpha_N)\}_{i=1}^w \subset \mathbf{PE}_{\leq d}(G,\mathcal{F}^{(1)})$$

and

$$M = \{M_i(\alpha_1,\cdots,\alpha_N)\}_{i=1}^w \subset \mathbf{PE}_{\leq d}(N)$$

having the property that for any $l-$cell partition

$$L \times M = \bigcup_{i=1}^{l} C_i,$$

there exist numbers a,b, and q, with $1 \leq a,b \leq w$ and $1 \leq q \leq l$, and sets $S_i \subset \{1,\cdots,N\}$, $1 \leq i \leq m$, with $S_1 < \cdots < S_m$, such that under the symbolic substitution $\beta_i = \bigcup_{n\in S_i} \alpha_n$, $1 \leq i \leq m$ we have, for $1 \leq i,j \leq t$,

$$\big(L_a(\alpha_1,\cdots,\alpha_N)R_i(\beta_1,\cdots,\beta_m),\ M_b(\alpha_1,\cdots,\alpha_N)W_j(\beta_1,\cdots,\beta_m)\big) \in C_q.$$

Proof. Let $N = mN(t^2,d,l)$ (where $N(t^2,d,l)$ is as in Theorem 2.10). If

$$F(\{1,\cdots,t^2\} \times \{1,\cdots,N\}^d) = \bigcup_{i=1}^{l} C_i$$

then one of the cells C_q of this partition, $1 \leq q \leq l$ contains a configuration of the form

$$\Big\{A \cup \big(B \times (S_1 \cup \cdots \cup S_m)^d\big) : B \subset \{1,\cdots,t^2\}\Big\},$$

where $A \subset (\{1,\cdots,t^2\} \times \{1,\cdots,N\}^d)$ and

$$\emptyset \neq S_j \subset \Big\{\frac{j-1}{m}N+1,\cdots,\frac{j}{m}N\Big\},\ \ 1 \leq j \leq m$$

satisfy

$$A\cap\big(\{1,\cdots,t^2\}\times(S_1\cup\cdots\cup S_m)^d\big)=\emptyset.$$

Now, for every fixed i,j, $1\le i,j\le t$, and $E\subset\{1,\cdots,N\}$ with $|E|\le d$, pick exactly one point $x_{E,i,j}\in(\{1,\cdots,N\})^d$ whose set of coordinates (a subset of $\{1,\cdots,N\}$) equals E. Write

$$E_z=E\cap\Big\{\frac{z-1}{m}N+1,\cdots,\frac{z}{m}N\Big\},\ \ 1\le z\le m,$$

and put $a_z=|E_z|$. Denote the elements of E_z by $n_z^{(1)}<\cdots<n_z^{(a_z)}$.

We wish to define a function

$$f:\ F\big(\{1,\cdots,t^2\}\times\{1,\cdots,N\}^d\big)\to\mathbf{PE}_{\le d}(N)\times\mathbf{PE}_{\le d}(N)$$

having some properties which we will put off mentioning for a bit and which satisfies $f(\alpha\cup\beta)=f(\alpha)f(\beta)$ when $\alpha\cap\beta=\emptyset$. Any such function will be uniquely determined once it is defined on singletons. For E,i,j as above, let

$$\begin{aligned}&f\big((it+j,x_{E,i,j})\big)\\&=\Big(R_i^{(a_1,\cdots,a_m)}(\alpha_{n_1^{(1)}},\cdots,\alpha_{n_1^{(a_1)}},\alpha_{n_2^{(1)}},\cdots,\alpha_{n_2^{(a_2)}},\cdots,\alpha_{n_m^{(1)}},\cdots,\alpha_{n_m^{(a_m)}}),\\&W_j^{(a_1,\cdots,a_m)}(\alpha_{n_1^{(1)}},\cdots,\alpha_{n_1^{(a_1)}},\alpha_{n_2^{(1)}},\cdots,\alpha_{n_2^{(a_2)}},\cdots,\alpha_{n_m^{(1)}},\cdots,\alpha_{n_m^{(a_m)}})\Big)\\&\quad\in\mathbf{PE}_{\le d}(N)\times\mathbf{PE}_{\le d}(N).\end{aligned}\tag{2.9}$$

We mean here, for example, by

$$R_i^{(a_1,\cdots,a_m)}(\alpha_{n_1^{(1)}},\cdots,\alpha_{n_1^{(a_1)}},\alpha_{n_2^{(1)}},\cdots,\alpha_{n_2^{(a_2)}},\cdots,\alpha_{n_m^{(1)}},\cdots,\alpha_{n_m^{(a_m)}}),$$

that element $S\in\mathbf{PE}_{\le d}(N)$ which is defined by

$$\begin{aligned}&S(\alpha_1,\cdots,\alpha_N)\\&=R_i^{(a_1,\cdots,a_m)}(\alpha_{n_1^{(1)}},\cdots,\alpha_{n_1^{(a_1)}},\alpha_{n_2^{(1)}},\cdots,\alpha_{n_2^{(a_2)}},\cdots,\alpha_{n_m^{(1)}},\cdots,\alpha_{n_m^{(a_m)}}).\end{aligned}\tag{2.10}$$

(This explains why we can say that the range of f is contained in $\mathbf{PE}_{\le d}(N)\times\mathbf{PE}_{\le d}(N)$.) On all other singletons in $F\big(\{1,\cdots,t^2\}\times\{1,\cdots,N\}^d\big)$, that is, on those not of the form $\big\{(it+j,x_{E,i,j})\big\}$, f is defined to be the identity. As mentioned earlier, f extends uniquely to a multiplicative function on $F\big(\{1,\cdots,t^2\}\times\{1,\cdots,N\}^d\big)$.

Let

$$L=\big\{L_i(\alpha_1,\cdots,\alpha_N):\ 1\le i\le w\big\}$$

be the set of elements in $\mathbf{PE}_{\le d}(N)$ which occur as the first coordinate in a member of the range of f. Recall that $\mathbf{PE}_{\le d}(G,\mathcal{F}^{(1)})$ is a group, and furthermore $(G,\mathcal{F}^{(1)})$ is assumed to be balanced (see Definition 2.11). Since $R_i\in\mathbf{PE}_{\le d}(G,\mathcal{F}^{(1)})$, $1\le i\le t$, $R_1^{(a_1,\cdots,a_m)}\in\mathbf{PE}_{\le d}(G,\mathcal{F}^{(1)})$ for $1\le i\le t$ and $a_i\in\mathbf{N}\cup\{0\}$, $1\le i\le m$. Therefore, by Proposition 2.4 the expression $S(\alpha_1,\cdots,\alpha_N)$ of 2.10 which occurs as the first coordinate in (2.9) is also in $\mathbf{PE}_{\le d}(G,\mathcal{F}^{(1)})$. Therefore we have $L\subset\mathbf{PE}_{\le d}(G,\mathcal{F}^{(1)})$. Let

$$M=\big\{M_i(\alpha_1,\cdots,\alpha_N):\ 1\le i\le w\big\}$$

be the set of elements which occur in the second coordinate. Of course the range of f is contained in $L \times M$. Suppose now that we have an l-cell partition of $L \times M$,

$$L \times M = \bigcup_{i=1}^{l} C_i.$$

Then

$$F(\{1, \cdots, t^2\} \times \{1, \cdots, N\}) = \bigcup_{i=1}^{l} f^{-1}(C_i),$$

and there exist q, $1 \leq q \leq m$, $A \subset \{1, \cdots, t^2\} \times \{1, \cdots, N\}$, and sets

$$\emptyset \neq S_j \subset \Big\{\frac{j-1}{m}N + 1, \cdots, \frac{j}{m}N\Big\}, \quad 1 \leq j \leq m,$$

where

$$A \cap \big(\{1, \cdots, t^2\} \times (S_1 \cup \cdots \cup S_m)^d\big) = \emptyset,$$

such that

$$\Big\{A \cup \big(B \times (S_1 \cup \cdots \cup S_m)^d\big) : B \subset \{1, \cdots, t^2\}\Big\} \subset f^{-1}(C_q).$$

Define a, b, $1 \leq a, b \leq w$ by $(L_a, M_b) = f(A)$, and let $\beta_i = \bigcup_{n \in S_i} \alpha_n$, $1 \leq i \leq m$. Now, as to the additional property of f, the reader should convince himself with the help of equation (2.8) that for $1 \leq i, j \leq t^2$,

$$f\big(it + j, (S_1 \cup \cdots \cup S_m)^d\big) = \big(R_i(\beta_1, \cdots, \beta_m), W_j(\beta_1, \cdots, \beta_m)\big).$$

It follows that

$$\big(L_a(\alpha_1, \cdots, \alpha_N)R_i(\beta_1, \cdots, \beta_m),\ M_b(\alpha_1, \cdots, \alpha_N)W_j(\beta_1, \cdots, \beta_m)\big) \in C_q,$$

$1 \leq i, j \leq t$.

□

A classical combinatorial result which concerns IP-rings was proved by Hindman in 1974. One formulation of Hindman's Theorem ([H]) is as follows:

Theorem 2.13 If $\mathcal{F}^{(1)}$ is an IP-ring, and if for some $r \in \mathbf{N}$, we are given an r-cell partition of $\mathcal{F}^{(1)}$, $\mathcal{F}^{(1)} = \bigcup_{i=1}^{r} C_i$, then for some i, $1 \leq i \leq r$, C_i contains an IP-ring $\mathcal{F}^{(2)}$.

A natural consequence of Hindman's Theorem is the following:

Proposition 2.14 (see [FK2], Theorem 1.5) Suppose that X is a compact metric space and that for each $n \in \mathbf{N}$, $\{x_\alpha^{(n)}\}_{\alpha \in \mathcal{F}}$ is an $\mathcal{F}$-sequence in X. Then for any IP-ring $\mathcal{F}^{(1)}$, there exists an IP-sub-ring $\mathcal{F}^{(2)} \subset \mathcal{F}^{(1)}$ such that

$$\underset{\alpha \in \mathcal{F}^{(2)}}{\text{IP-lim}}\ x_\alpha^{(n)} = z_n$$

exists for each $n \in \mathbf{N}$.

We will find it convenient to use the following generalization of Hindman's Theorem, the Milliken-Taylor Theorem.

Theorem 2.15 ([M], [T]) Suppose that $\mathcal{F}^{(1)}$ is an IP-ring, $m, r \in \mathbf{N}$, and $(\mathcal{F}^{(1)})^m_< = \bigcup_{i=1}^r C_i$. Then there exists j, $1 \leq i \leq r$, and an IP-ring $\mathcal{F}^{(2)} \subset \mathcal{F}^{(1)}$ such that $(\mathcal{F}^{(2)})^m_< \subset C_j$.

Just as Hindman's Theorem gives rise to Proposition 2.14, the following is a consequence of the Milliken-Taylor Theorem:

Theorem 2.16 Suppose that X is a compact metric space, $(m_n)_{n\in\mathbf{N}} \subset \mathbf{N}$ is a sequence, and that for each $n \in \mathbf{N}$, $\{x^{(n)}_{(\alpha_1,\cdots,\alpha_{m_n})} : (\alpha_1, \cdots, \alpha_{m_n}) \in (\mathcal{F})^{m_n}_<\}$ is a sequence in X indexed by $(\mathcal{F})^{m_n}_<$. Then, for any IP-ring $\mathcal{F}^{(1)}$, there exists an IP-sub-ring $\mathcal{F}^{(2)} \subset \mathcal{F}^{(1)}$ such that

$$\operatorname*{IP-lim}_{(\alpha_1,\cdots,\alpha_{m_n})\in(\mathcal{F}^{(2)})^{m_n}_<} x^{(n)}_{(\alpha_1,\cdots,\alpha_{m_n})} = z_n$$

exists for each $n \in \mathbf{N}$.

The following theorem from [BFM] will be used in Section 3 in order to establish the existence of primitive extensions.

Theorem 2.17 Suppose that $\mathcal{H}$ is a Hilbert space, $(U_i)_{i=1}^t$ is a commuting family of unitary operators on $\mathcal{H}$, $\big(p_i(x_1, \cdots, x_k)\big)_{i=1}^t \subset \mathbf{Z}[x_1, \cdots, x_k]$, $p_i(0, \cdots, 0) = 0$, $1 \leq i \leq t$, and that $(n^{(j)}_\alpha)_{\alpha\in\mathcal{F}}$ are IP-sets in $\mathbf{Z}$, $1 \leq j \leq k$. Suppose $\mathcal{F}^{(1)}$ is an IP-ring such that for each $f \in \mathcal{H}$,

$$\operatorname*{IP-lim}_{\alpha\in\mathcal{F}^{(1)}} \Big(\prod_{i=1}^t U_i^{p_i(n^{(1)}_\alpha,\cdots,n^{(k)}_\alpha)}\Big) f = P_{(p_1,\cdots,p_t)} f$$

exists in the weak topology. Then $P_{(p_1,\cdots,p_t)}$ is an orthogonal projection. Projections of this type commute, that is, if also $\big(q_i(x_1, \cdots, x_k)\big)_{i=1}^t \subset \mathbf{Z}[x_1, \cdots, x_k]$, $q_i(0, \cdots, 0) = 0$, $1 \leq i \leq t$, then $P_{(p_1,\cdots,p_t)} P_{(q_1,\cdots,q_t)} = P_{(q_1,\cdots,q_t)} P_{(p_1,\cdots,p_t)}$.

We shall need one more proposition which, while not being as purely "combinatorial" in nature as the others in this section, is concerned with limits along IP-rings, in particular providing a sufficient condition for showing that an $\mathcal{F}$-sequence in a Hilbert space converges to 0 weakly along a subring. This proposition is a modification of [FK2, Lemma 5.3]. A proof is provided for completeness.

Proposition 2.18 Suppose that $(x_\alpha)_{\alpha\in\mathcal{F}}$ is a bounded $\mathcal{F}$-sequence in a Hilbert space and $\mathcal{F}^{(1)}$ is an IP-ring. If

$$\operatorname*{IP-lim}_{(\beta,\alpha)\in(\mathcal{F}^{(1)})^2_<} \langle x_\alpha, x_{\alpha\cup\beta}\rangle = 0$$

then for some subring $\mathcal{F}^{(2)} \subset \mathcal{F}^{(1)}$,

$$\operatorname*{IP-lim}_{\alpha\in\mathcal{F}^{(2)}} x_\alpha = 0$$

in the weak topology.

Proof. Let $\mathcal{F}^{(2)} \subset \mathcal{F}^{(1)}$ be an IP-ring with the property that

$$\operatorname*{IP-lim}_{\alpha\in\mathcal{F}^{(2)}} x_\alpha = u$$

exists weakly. We have, for all $k \in \mathbf{N}$,

$$\text{IP-}\lim_{(\alpha_1,\cdots,\alpha_k)\in(\mathcal{F}^{(2)})^k_<} \langle x_{\alpha_1\cup\cdots\cup\alpha_k}, u\rangle = ||v||^2,$$

from which it follows that for all $m \in \mathbf{N}$,

$$\text{IP-}\lim_{(\alpha_1,\cdots,\alpha_m)\in(\mathcal{F}^{(2)})^m_<} \Big\langle \frac{1}{m}\sum_{k=1}^m x_{\alpha_1\cup\cdots\cup\alpha_k}, u\Big\rangle = ||v||^2. \tag{2.11}$$

On the other hand,

$$\begin{aligned}
&\lim_{m\to\infty} \text{IP-}\lim_{(\alpha_1,\cdots,\alpha_m)\in(\mathcal{F}^{(2)})^m_<} \Big|\Big|\frac{1}{m}\sum_{k=1}^m x_{\alpha_1\cup\cdots\cup\alpha_k}\Big|\Big|^2 \\
&= \lim_{m\to\infty} \text{IP-}\lim_{(\alpha_1,\cdots,\alpha_m)\in(\mathcal{F}^{(2)})^m_<} \frac{1}{m^2}\sum_{k,j=1}^m \langle x_{\alpha_1\cup\cdots\cup\alpha_k}, x_{\alpha_1\cup\cdots\cup\alpha_j}\rangle \\
&= \lim_{m\to\infty} \frac{1}{m^2}\sum_{k,j=1}^m \text{IP-}\lim_{(\alpha_1,\cdots,\alpha_m)\in(\mathcal{F}^{(2)})^m_<} \langle x_{\alpha_1\cup\cdots\cup\alpha_k}, x_{\alpha_1\cup\cdots\cup\alpha_j}\rangle \\
&= \lim_{m\to\infty} \frac{1}{m^2}\sum_{k=1}^m ||x_{\alpha_1\cup\cdots\cup\alpha_k}||^2 = 0.
\end{aligned}$$

This together with (2.11) gives $u = 0$.

2.3. Factorization and Decomposition of Measures.

We conclude this section with a brief review of generalities concerning factors and measurable disintegration. For a more comprehensive treatment, the reader is referred to [F2].

We will be dealing with measure preserving systems on probability spaces, denoted by $(Y, \mathcal{B}, \nu, \Omega)$, etc. in which the group Ω of measure preserving transformations is generated by r commuting transformations $T_1, T_2, \cdots, T_r$. We will not usually need to explicitly mention the transformations T_i, rather we deal with elements of $\mathbf{PE}(m)$, which we denote by $T(\alpha_1, \cdots, \alpha_m)$, $S(\alpha_1, \cdots, \alpha_m)$, etc. We have already indicated these may be viewed as $\mathcal{F}$-sequences in Ω.

Remark 2.19 For the purposes of the proofs of our theorems, we shall always assume (implicitly) that the measure spaces with which we deal are *Lebesgue spaces.* In other words, we shall assume they are measurably isomorphic to a union of at most countably many (possibly zero) point masses and some (possibly trivial) interval on the real line endowed with Lebesgue measure. This is done because many of the constructions we use (decomposition of measures, etc.) require the underlying spaces to be Lebesgue. Later we will explain why our results hold for general, possibly non-Lebesgue spaces, given that they hold for Lebesgue spaces (Remark 5.4).

At times we will be dealing with a system $(Y, \mathcal{B}, \nu, \Omega)$ which is a *factor* of some other system, say $(Z, \mathcal{C}, \gamma, \Omega)$. In other words, we have measure preserving actions of the same group Ω on both spaces, and moreover there exists a measurable function $\pi : Z \to Y$, called the *factor map*, such that $\gamma(\pi^{-1}(B)) = \nu(B)$ for all $B \in \mathcal{B}$ and such that $\pi(Tz) = T\pi(z)$ for a.e. $z \in Z$ and every $T \in \Omega$.

Suppose that $(Y, \mathcal{B}, \nu, \Omega)$ is a factor of $(Z, \mathcal{C}, \gamma, \Omega)$ and π is the factor map. Then of course $Z = \bigcup_{y \in Y} \pi^{-1}(y)$ mod 0. We will identify $\mathcal{B}$ with the completion of the σ-algebra $\{\bigcup_{y \in B} \pi^{-1}(y) : B \in \mathcal{B}\}$ of subsets of Z which consists of all sets which "sit over" measurable subsets of Y. Therefore we will regularly say "$C \in \mathcal{B}$" if $C \subset Z$ and $C = \pi^{-1}(B)$ mod 0 for some $B \in \mathcal{B}$. (In other words, we regard $\mathcal{B}$ as a sub-σ-algebra of $\mathcal{C}$.)

There exists a family $\{\gamma_y : y \in Y\}$ of probability measures on Z, called the decomposition of the measure γ over Y, having the property that for every $f \in L^1(Z, \mathcal{C}, \gamma)$, the function

$$E(f|Y)(y) = E(f|\mathcal{B})(y) = \int_Z f(z)\, d\gamma_y(z),$$

called the *conditional expectation* of f given $\mathcal{B}$, is defined a.e., is a measurable function of y, and satisfies $\int_{\pi^{-1}(B)} f\, d\gamma = \int_B E(f|Y)\, d\nu$ for all $B \in \mathcal{B}$. Sometimes we will write γ_z, where $z \in Z$. In this case, we mean $\gamma_{\pi(z)}$.

The conditional product system of Z with itself relative to Y will be denoted by $(Z \times_Y Z, \mathcal{C} \otimes_{\mathcal{B}} \mathcal{C}, \tilde{\gamma}, \tilde{\Omega})$. This space arises in the following way. For $f, g \in L^\infty(Z, \mathcal{C}, \gamma)$, write $f \otimes g(z_1, z_2) = f(z_1)g(z_2)$. Define a measure $\tilde{\gamma}$ on $\mathcal{C} \otimes \mathcal{C}$ by letting

$$\begin{aligned} \int f \otimes g\, d\tilde{\gamma} &= \int_Y E(f|Y)(y) E(g|Y)(y)\, d\nu(y) \\ &= \int_Y \int_Z \int_Z f(z_1) g(z_2)\, d\gamma_y(z_1) d\gamma_y(z_2) d\nu(y) \end{aligned}$$

for $f, g \in L^\infty(Z, \mathcal{C}, \gamma)$ and extending to $\mathcal{C} \otimes \mathcal{C}$. If we let $\pi : Z \to Y$ denote the natural factor map, one may show that the measure $\tilde{\gamma}$ is supported on the set

$$Z \times_Y Z = \big\{(z_1, z_2) \in Z \times Z : \pi(z_1) = \pi(z_2)\big\}.$$

We denote by $\mathcal{C} \otimes_{\mathcal{B}} \mathcal{C}$ the σ-algebra on $Z \times_Y Z$ which consists of the members of $\mathcal{C} \otimes \mathcal{C}$ intersected with $Z \times_Y Z$. The group of $\tilde{\gamma}$-preserving transformations $\tilde{\Omega}$ is generated by the set of r commuting measure preserving transformations $\{\tilde{T}_1, \cdots, \tilde{T}_r\}$, where $\tilde{T}_i$ is defined by $\tilde{T}_i(z_1, z_2) = (T_i z_1, T_i z_2)$, $1 \le i \le r$. If $T \in \mathbf{PE}(1)$ is given by $T(\alpha) = \prod_{i=1}^r T_i^{p_i(n_\alpha^{(1)}, \cdots, n_\alpha^{(k)})}$ then we will write $\tilde{T}$ for the $\mathcal{F}$-sequence in $\tilde{\Omega}$ defined by

$$\tilde{T}(\alpha) = \prod_{i=1}^r \tilde{T}_i^{p_i(n_\alpha^{(1)}, \cdots, n_\alpha^{(k)})}.$$

When there is not cause for confusion, however (i.e. when it is clear that the sequence of transformations lies in $\tilde{\Omega}$), we may simply use T instead of $\tilde{T}$. (See for example Definitions 3.5 and 3.6, where this is the case.)

CHAPTER 3

PRIMITIVE EXTENSIONS

Recall that (see the discussion near the end of the introduction) the crucial step in the proof of our main result is showing that if everything holds for sets in some proper factor of our system then it holds for sets in some non-trivial extension of this factor. The non-trivial extensions we will be finding for which it is possible to prove this fact are so-called *primitive* extensions. Loosely speaking we call an extension Z of Y *primitive* if for some $d \in \mathbf{N}$ there exists a subgroup $G < \mathbf{PE}_{\leq d}(1)$ having the property that Z-measurable functions exhibit certain *rigidity* properties along members of G on an appropriately chosen IP-ring, while exhibiting certain *mixing* properties along members of $\mathbf{PE}_{\leq d}(1) \setminus G$. This of course is a rather imprecise explanation and the intention is merely to offer some of the flavor of this section.

The following definition makes somewhat more precise what we mean by "rigid" behavior. It is modelled after Definition 6.1 of [FK2], although it is somewhat more cumbersome than the definition appearing there. Our definition of course takes into account polynomiality, but even in the linear case it is weaker than Definition 6.1 in [FK2] and applies to a possibly larger class of functions. Our reason for choosing the present formulation is that at some point we will want to show the existence of *almost periodic* functions (see Theorem 3.11), and this is more easily accomplished with the weakened definition, which is still strong enough for our later purposes. The corresponding portion of the proof of Furstenberg's and Katznelson's linear IP multiple recurrence theorem, namely Lemmas 7.1, 7.2 and 7.3 of [FK2], uses linearity of the expressions, and while that proof could probably be adapted to the polynomialized situation, we are choosing to take an easier road home, sacrificing a bit of simplicity in the definitions in order to shorten the proof.

We remind the reader that all measure spaces we work with are assumed to be Lebesgue spaces.

Definition 3.1 Suppose that $(Y, \mathcal{B}, \nu, \Omega)$ is a factor of $(Z, \mathcal{C}, \gamma, \Omega)$, $d \in \mathbf{N}$, $G \subset \mathbf{PE}_{\leq d}(1)$ is a subgroup, and $\mathcal{F}^{(1)}$ is an IP-ring. We say that $f \in \mathrm{L}^2(Z, \mathcal{C}, \gamma)$ is *almost periodic over* Y *for* G *along* $\mathcal{F}^{(1)}$, and we write $f \in AP(Z, Y, G, \mathcal{F}^{(1)})$, if for every $\epsilon > 0$ there exist $g_1, \cdots, g_l \in \mathrm{L}^2(Z, \mathcal{C}, \gamma)$ and $D \in \mathcal{B}$ with $\nu(D) < \epsilon$ such that for every $\delta > 0$ and every $T(\alpha_1, \cdots, \alpha_m) \in \mathbf{PE}(G, \mathcal{F}^{(1)})$ there exists $\alpha_0 \in \mathcal{F}^{(1)}$ having the property that whenever $(\alpha_1, \cdots, \alpha_m) \in (\mathcal{F}^{(1)})^m_<$ with $\alpha_0 < \alpha_1$ we have a set $E = E(\alpha_1, \cdots, \alpha_m) \in \mathcal{B}$, with $\nu(E) < \delta$, such that for all $y \notin (D \cup E)$, there exists $j = j(\alpha_1, \cdots, \alpha_m, y)$ with $1 \leq j \leq l$ satisfying

$$\left|\left|T(\alpha_1, \cdots, \alpha_m)f - g_j\right|\right|_y < \epsilon.$$

We realize that this definition is quite involved. However, it strikes a balance between being both strong enough to include the properties we desire and weak enough to simplify the task of showing that almost periodic functions are abundant under the proper conditions (this is the content of Theorem 3.11).

Remarks 3.2

(i) $AP(Z,Y,G,\mathcal{F}^{(1)}) \cap L^\infty(Z,\mathcal{C},\gamma)$ is a T-invariant algebra which contains $|g|$ and $\overline{g}$ whenever it contains g.

(ii) If $f \in \mathrm{L}^2(Z,\mathcal{C},\gamma)$ and for every $\epsilon > 0$ there exists $h \in AP(Z,Y,G,\mathcal{F}^{(1)})$ with $\left|\left|f-h\right|\right|_y < \epsilon$ for a.e. $y \in Y$, then $f \in AP(Z,Y,G,\mathcal{F}^{(1)})$.

(iii) If $h \in AP(Z,Y,G,\mathcal{F}^{(1)})$ and $E \in \mathcal{B}$ then $1_E h \in AP(Z,Y,G,\mathcal{F}^{(1)})$.

(iv) $AP(Z,Y,G,\mathcal{F}^{(1)})$ need not be closed; however, if $f \in \overline{AP(Z,Y,G,\mathcal{F}^{(1)})}$ then for any $\epsilon > 0$ there exist $g_1,\cdots,g_l \in L^\infty(Z,\mathcal{C},\gamma)$ (select them first in $L^2(Z,\mathcal{C},\gamma)$ and then truncate) such that for every $T(\alpha_1,\cdots,\alpha_m) \in \mathbf{PE}(G,\mathcal{F}^{(1)})$ there exists $\alpha_0 \in \mathcal{F}^{(1)}$ having the property that for every $(\alpha_1,\cdots,\alpha_m) \in (\mathcal{F}^{(1)})^m_<$ with $\alpha_0 < \alpha_1$ we have a set $E = E(\alpha_1,\cdots,\alpha_m) \in \mathcal{B}$ with $\nu(E) < \epsilon$ such that for all $y \in E^c$ there exists $j = j(\alpha_1,\cdots,\alpha_m,y)$ with $1 \le j \le l$ satisfying $||T(\alpha_1,\cdots,\alpha_m)f - g_j||_y < \epsilon$. (Notice the absence of D and δ in this weaker property.)

The following theorem is the first of three theorems in this section which will be used in the main proof (Section 5). Basically it allows us to assume without loss of generality that, subject to constraints which we are able to ensure are satisfied there, 1_A is almost periodic.

Theorem 3.3 Suppose that $1_A \in \overline{AP(Z,Y,G,\mathcal{F}^{(1)})}$ (closure in $\mathrm{L}^2(Z,\mathcal{C},\gamma)$) and $\delta > 0$. Then there exists a set $A' \subset A$ with $\mu(A \setminus A') < \delta$ such that $1_{A'} \in AP(Z,Y,G,\mathcal{F}^{(1)})$.

Proof. For each $\epsilon > 0$, we can find $h \in AP(Z,Y,G,\mathcal{F}^{(1)})$ and a $\mathcal{B}$-measurable set E with $\nu(E) > 1-\epsilon$ such that for each $y \in E$, $\left|\left|1_A - h\right|\right|_y < \epsilon$. We have, by Remark 3.2 (iii), $1_E h \in AP(Z,Y,G,\mathcal{F}^{(1)})$, and furthermore $\left|\left|1_{A\cap E} - 1_E h\right|\right|_y < \epsilon$ for a.e. $y \in Y$. Replace A by $A \cap E$. Repeat this process countably, for a sequence of ϵ's whose sum is less than δ. We are left with a set A' which, by construction and Remark 3.2 (ii), has the properties we require.

□

Our next task is to define primitive extensions. First we need two preliminary definitions. The first of these comes nearly verbatim from [FK2].

Definition 3.4 If $(Y,\mathcal{B},\nu)$ is a factor of $(Z,\mathcal{C},\gamma)$, then a function $H(z_1,z_2) \in L^\infty(Z\times_Y Z, \mathcal{C}\otimes_{\mathcal{B}}\mathcal{C},\tilde{\gamma})$ is called a $\frac{Z}{Y}$-kernel if

$$\int H(z_1,z_2)d\gamma_{z_2}(z_1) = 0$$

for a.e. $z_2 \in Z$. For $\frac{Z}{Y}$-kernels H we define, for $\phi \in \mathrm{L}^2(Z,\mathcal{C},\gamma)$,

$$H * \phi(z_1) = \int H(z_1,z_2)\phi(z_2)d\gamma_{z_1}(z_2).$$

If H is a $\frac{Z}{Y}$-kernel, then for a.e. $y \in Y$, $\phi \to H * \phi$ defines a Hilbert-Schmidt operator on $\mathrm{L}^2(Z,\mathcal{C},\gamma_y)$. If H is self-adjoint, that is if $H(z_1,z_2) = \overline{H(z_2,z_1)}$ a.e, then for a.e. $y \in Y$, $\phi \to H * \phi$ defines a self-adjoint Hilbert-Schmidt operator on

$L^2(Z, \mathcal{C}, \gamma_y)$ with real spectrum $\big(\lambda_i(y)\big)_{i=1}^{\infty}$, $|\lambda_1(y)| \geq |\lambda_2(y)| \geq \cdots$. Also, we may write

$$H(z_1, z_2) = \sum_k \lambda_k(y)\phi_k(z_1)\overline{\phi_k(z_2)},$$

where $\{\phi_k\}_{k=1}^{\infty} \subset L^2(Z, \mathcal{C}, \gamma)$ is, for a.e. $y \in Y$, an orthonormal (in $L^2(Z, \mathcal{C}, \gamma_y)$) family of eigenvectors for this operator associated with the eigenvalues $\lambda_k(y)$ (see [FK2] for details). If for all k we have $\lambda_k \geq 0$ a.e., H is said to be *non-negative definite.*

As mentioned earlier, there are two components to primitive extensions; rigidity, which is exemplified in the phenomenon of almost periodicity, and *mixing*, which we now address.

Definition 3.5 Suppose that $\mathcal{F}^{(1)}$ is an IP-ring, $(Y, \mathcal{B}, \nu, \Omega)$ is a factor of $(Z, \mathcal{C}, \gamma, \Omega)$, and $S \in \mathbf{PE}(1)$. S is said to be *mixing on Z relative to Y along $\mathcal{F}^{(1)}$* if for every $H \in L^2(Z \times_Y Z, \mathcal{C} \otimes_{\mathcal{B}} \mathcal{C}, \tilde{\gamma})$ satisfying $E(H|Y) = 0$ we have

$$\operatorname*{IP-lim}_{\alpha \in \mathcal{F}^{(1)}} S(\alpha)H = 0$$

weakly. A set $A = \{S_1, S_2, \cdots, S_t\} \subset \mathbf{PE}(1)$ will be said to be a *mixing set* on Z relative to Y along $\mathcal{F}^{(1)}$ if S_i and $S_i S_j^{-1}$ are mixing on Z relative to Y along $\mathcal{F}^{(1)}$, $1 \leq i \neq j \leq t$.

Finally we are in a position to define primitive extensions. We point out that our approach to this matter is somewhat different from that in [FK2]. There, the density of almost periodic functions is taken as part of the definition of primitive extension and the bulk of the work is showing that a particular class of extensions (namely, those arising in a prescribed manner) have this property. We, on the other hand, start by defining an extension to be primitive if it arises in this (suitably altered because of polynomiality, of course) manner, and prove the density of almost periodic functions (for this prescribed class) as Theorem 3.11. The effect is that the task of showing that primitive extensions exist (Theorem 3.15) is correspondingly simplified. The difference between these two approaches, we remark, is more ornamental than fundamental.

Definition 3.6 Suppose that $(Y, \mathcal{B}, \nu, \Omega)$ is a factor of $(W, \mathcal{D}, \xi, \Omega)$, $d \in \mathbf{N}$, $G < \mathbf{PE}_{\leq d}(1)$ is a subgroup, $\mathcal{F}^{(1)}$ is an IP-ring, and there exists a non-trivial self-adjoint non-negative definite $\frac{W}{Y}$-kernel H such that for all $T \in G$ we have

$$\operatorname*{IP-lim}_{\alpha \in \mathcal{F}^{(1)}} T(\alpha)H = H.$$

Suppose an Ω-invariant σ-algebra $\mathcal{C} \subset \mathcal{W}$ is complete with regard to ξ and is contained in the least Ω-invariant σ-algebra containing $\mathcal{B}$ with respect to which the functions

$$\{H * \phi : \ \phi \in L^2(W, \mathcal{D}, \xi)\}$$

are measurable. In this case, the factor determined by $\mathcal{C}$, say $(Z, \mathcal{C}, \gamma, \Omega)$, is said to be a *compact extension of Y with respect to G along $\mathcal{F}^{(1)}$*. Suppose in addition that T is mixing on Z relative to Y along $\mathcal{F}^{(1)}$ for all $T \in (\mathbf{PE}_{\leq d}(1) \setminus G)$. In this case $(Z, \mathcal{C}, \gamma, \Omega)$ is said to be a *primitive extension of Y along $\mathcal{F}^{(1)}$ with compact part G.*

Notice that the definition does depend in a critical way on d, so that it would be perhaps more precise to say "d-primitive extension". However, in Section 5, where these notions are applied, d will be fixed and there won't be cause for confusion.

Remarks 3.7

(i) If H (the H of Definition 3.6, that is) is given by

$$H(w_1, w_2) = \sum_k \lambda_k(y)\phi_k(w_1)\overline{\phi_k(w_2)},$$

then for w_1 in the fiber over y we have:

$$H * \phi_j(w_1) = \int \Big(\sum_k \lambda_k(y)\phi_k(w_1)\overline{\phi}_k(w_2)\Big)\phi_j(w_2)\, d\xi_{w_1}(w_2) = \lambda_j(y)\phi_j(w_1),$$

where $\{\xi_w : w \in W\}$ represents the disintegration of ξ over the factor Y. It follows that the functions ϕ_j must be Z-measurable. In particular, H is a $\frac{Z}{Y}$-kernel. Moreover, if $\phi \in L^2(W, \mathcal{D}, \xi)$ and $\phi' = E(\phi|Z)$ then $\langle \phi, \phi_k\rangle_w = \langle \phi', \phi_k\rangle_w$ a.e., so that

$$\begin{aligned} H * \phi(w_1) &= \int \Big(\sum_k \lambda_k(y)\phi_k(w_1)\overline{\phi_k(w_2)}\Big)\phi(w_2)\, d\xi_{w_1}(w_2) \\ &= \sum_k \lambda_k(y)\phi_k(w_1)\langle \phi, \phi_k\rangle_{w_1} \\ &= \sum_k \lambda_k(y)\phi_k(w_1)\langle \phi', \phi_k\rangle_{w_1} = H * \phi'(w_1). \end{aligned}$$

Hence $\mathcal{C}$ must be contained in the least Ω-invariant σ-algebra containing $\mathcal{B}$ with respect to which $H * \phi$ is measurable for all $\phi \in L^2(Z, \mathcal{C}, \gamma)$. But $\{H * \phi : \phi \in L^2(Z, \mathcal{C}, \gamma)\}$ is an algebra. Hence

$$L^2(Z, \mathcal{C}, \gamma) = \overline{\{H * \phi : \phi \in L^2(Z, \mathcal{C}, \gamma)\}}. \tag{3.1}$$

(ii) We now follow the construction of Definition 3.6 with $W = Z \times_Y Z$. Consider the $\frac{Z \times_Y Z}{Y}$-kernel $H' \in \mathrm{L}^2\big((Z \times_Y Z) \times_Y (Z \times_Y Z)\big)$ given by

$$H'(z_1, z_2, z_3, z_4) = \sum_{k,l} \lambda_k(y)\lambda_l(y)\phi_k(z_1)\phi_l(z_2)\overline{\phi_k(z_3)\phi_l(z_4)}.$$

Then

$$\operatorname*{IP-lim}_{\alpha \in \mathcal{F}^{(1)}} T(\alpha)H' = H'$$

for all $T \in G$. Furthermore,

$$\{H' * \phi :\ \phi \in \mathrm{L}^2(Z \times_Y Z)\}$$

is dense in $\mathrm{L}^2(Z \times_Y Z, \mathcal{C} \otimes_{\mathcal{B}} \mathcal{C}, \tilde{\gamma})$, so that $Z \times_Y Z$ is a compact extension of Y with respect to G along $\mathcal{F}^{(1)}$.

(iii) If $T \in \mathbf{PE}(1)$ is mixing on Z relative to Y along $\mathcal{F}^{(1)}$ then for any $H_1, H_2 \in \mathrm{L}^2(Z \times_Y Z)$, with either $E(H_1|Y) = 0$ or $E(H_2|Y) = 0$, the function

$$H(z_1, z_2, z_3, z_4) = H_1(z_1, z_2)H_2(z_3, z_4) \in \mathrm{L}^2\big((Z \times_Y Z) \times_Y (Z \times_Y Z)\big)$$

satisfies

$$\operatorname*{IP-lim}_{\alpha \in \mathcal{F}^{(1)}} T(\alpha)H = 0$$

weakly. As linear combinations of such functions H are dense in

$$\{H \in \mathrm{L}^2\big((Z \times_Y Z) \times_Y (Z \times_Y Z)\big) : E(H|Y) = 0\},$$

T is mixing on $Z \times_Y Z$ relative to Y along $\mathcal{F}^{(1)}$.

In light of Remarks 3.7, we have the following.

Proposition 3.8 If $(Z, \mathcal{C}, \gamma, \Omega)$ is a primitive extension of $(Y, \mathcal{B}, \nu, \Omega)$ along $\mathcal{F}^{(1)}$ with compact part G, then $(Z \times_Y Z, \mathcal{C} \otimes_{\mathcal{B}} \mathcal{C}, \tilde{\gamma}, \tilde{\Omega})$ is also a primitive extension of $(Y, \mathcal{B}, \nu, \Omega)$ along $\mathcal{F}^{(1)}$ with compact part G.

We now turn our attention to showing (Theorem 3.11) that if $(Z, \mathcal{C}, \gamma, \Omega)$ is a compact extension of $(Y, \mathcal{B}, \nu, \Omega)$ relative to G along $\mathcal{F}^{(1)}$ then for some *refinement* (i.e. subring) $\mathcal{F}^{(2)} \subset \mathcal{F}^{(1)}$, $AP(Z, Y, G, \mathcal{F}^{(2)})$ is dense in $\mathrm{L}^2(Z, \mathcal{C}, \gamma)$. The following proposition is the key.

Proposition 3.9 Let $(Z, \mathcal{C}, \gamma, \Omega)$ be an extension of $(Y, \mathcal{B}, \nu, \Omega)$. If $d \in \mathbf{N}$, $G < \mathbf{PE}_{\leq d}(1)$ is a subgroup, $\mathcal{F}^{(1)}$ is an IP-ring and there exists a $\frac{W}{Y}$-kernel H such that for all $T \in G$ we have

$$\operatorname*{IP-lim}_{\alpha \in \mathcal{F}^{(1)}} T(\alpha)H = H,$$

then there exists a refinement $\mathcal{F}^{(2)} \subset \mathcal{F}^{(1)}$ such that for all $T(\alpha_1, \cdots, \alpha_m) \in \mathbf{PE}(G, \mathcal{F}^{(2)})$ and all $\epsilon > 0$ there exists $\alpha_0 \in \mathcal{F}^{(2)}$ far enough out that whenever $(\alpha_1, \cdots, \alpha_m) \in (\mathcal{F}^{(2)})^m_<$ with $\alpha_0 < \alpha_1$ we have

$$\big|\big|T(\alpha_1, \cdots, \alpha_m)H - H\big|\big| < \epsilon.$$

Proof. By Theorem 2.14, we may choose a refinement $\mathcal{F}^{(2)} \subset \mathcal{F}^{(1)}$ with the property that for every polynomial expression $T(\alpha_1, \cdots, \alpha_m) \in \mathbf{PE}(m)$, we have existence of the limit

$$\operatorname*{IP-lim}_{(\alpha_1, \cdots, \alpha_m) \in (\mathcal{F}^{(1)})^m_<} \big|\big|T(\alpha_1, \cdots, \alpha_m)H - H\big|\big|.$$

We need only show that, for arbitrary $T(\alpha_1, \cdots, \alpha_m) \in \mathbf{PE}(G, \mathcal{F}^{(2)})$ (which we now fix), this limit is zero. To do this, it suffices to find, for arbitrary $\epsilon > 0$ and an arbitrary refinement $\mathcal{F}^{(3)} \subset \mathcal{F}^{(2)}$ (both of which we now fix), some $(\alpha_1, \cdots, \alpha_m) \in (\mathcal{F}^{(3)})^m_<$ with

$$\big|\big|T(\alpha_1, \cdots, \alpha_m)H - H\big|\big| < \epsilon.$$

Recall that we have

$$T(\alpha_1, \cdots, \alpha_m) = S(\alpha_1)S^{(\alpha_1)}(\alpha_2) \cdots S^{(\alpha_1, \cdots, \alpha_{m-1})}(\alpha_m),$$

where $S \in G$, and where for some $\alpha_0 \in \mathcal{F}^{(3)}$ we have, for any fixed

$$(\alpha_1, \cdots, \alpha_{m-1}) \in (\mathcal{F}^{(3)})^{m-1}_<$$

with $\alpha_1 > \alpha_0$,

$$\big\{S^{(\alpha_1)}, \cdots, S^{(\alpha_1, \cdots, \alpha_{m-1})}\big\} \subset G.$$

As $S \in G$, we may fix $\alpha_1 > \alpha_0$ with

$$||S(\alpha_1)H - H|| < \frac{\epsilon}{m}.$$

As now $S^{(\alpha_1)} \in G$, we may fix $\alpha_2 \in \mathcal{F}^{(3)}$ with $\alpha_2 > \alpha_1$ such that

$$||S^{(\alpha_1)}(\alpha_2)H - H|| < \frac{\epsilon}{m}.$$

As $S^{(\alpha_1,\alpha_2)} \in G$, we may fix α_3, etc. Finally, we have $\alpha_m \in \mathcal{F}^{(3)}$, $\alpha_m > \alpha_{m-1}$, with

$$||S^{(\alpha_1,\cdots,\alpha_{m-1})}(\alpha_m)H - H|| < \frac{\epsilon}{m}.$$

Now we have

$$\begin{aligned}
&||T(\alpha_1,\cdots,\alpha_m)H - H|| \\
&= ||S(\alpha_1)S^{(\alpha_1)}(\alpha_2)\cdots S^{(\alpha_1,\cdots,\alpha_{m-1})}(\alpha_m)H - H|| \\
&\le ||S(\alpha_1)S^{(\alpha_1)}(\alpha_2)\cdots S^{(\alpha_1,\cdots,\alpha_{m-1})}(\alpha_m)H - S^{(\alpha_1)}(\alpha_2)\cdots S^{(\alpha_1,\cdots,\alpha_{m-1})}(\alpha_m)H|| \\
&\quad + ||S^{(\alpha_1)}(\alpha_2)\cdots S^{(\alpha_1,\cdots,\alpha_{m-1})}(\alpha_m)H - S^{(\alpha_1,\alpha_2)}(\alpha_3)\cdots S^{(\alpha_1,\cdots,\alpha_{m-1})}(\alpha_m)H|| \\
&\quad + \cdots + ||S^{(\alpha_1,\cdots,\alpha_{m-1})}(\alpha_m)H - H|| \\
&= ||S(\alpha_1)H - H|| + ||S^{(\alpha_1)}(\alpha_2)H - H|| + \cdots + ||S^{(\alpha_1,\cdots,\alpha_{m-1})}(\alpha_m)H - H|| < \epsilon.
\end{aligned}$$

□

The content of the following lemma is more or less a relativized extension of the fact that the range over a bounded set in a Hilbert space of the operator induced by a bounded kernel is pre-compact. It is needed for Theorem 3.11.

Lemma 3.10 Suppose that $(Z,\mathcal{C},\gamma,\Omega)$ is an extension of $(Y,\mathcal{B},\nu,\Omega)$, H is a $\frac{Z}{Y}$-kernel, and $\epsilon > 0$. There exist functions $g_1,\cdots,g_l \in \mathrm{L}^2(Z,\mathcal{C},\gamma)$ and a set $D \in \mathcal{B}$ with $\nu(D) < \epsilon$ having the property that for every $f \in L^\infty(Z,\mathcal{C},\gamma)$ with $||f||_\infty < 1$ and every $y \in D^c$, $||H * f - g_j||_y < \epsilon$ for some j, $1 \le j \le l$.

Proof. We may assume without loss of generality that $||H||_\infty \le 1$. Let

$$B_y = \{f \in \mathrm{L}^2(Z,\mathcal{C},\gamma_y) : \ ||f||_y < 1\}.$$

Let $(h_i)_{i=1}^\infty$ be a sequence of functions dense in $\mathrm{L}^2(Z,\mathcal{C},\gamma)$ (therefore dense in $\mathrm{L}^2(Z,\mathcal{C},\gamma_y)$ for a.e. $y \in Y$). Then $(H * h_i)_{i=1}^\infty$ is dense in $H * B_y$ for a.e. y. But $H * B_y$ is totally bounded in $\mathrm{L}^2(Z,\mathcal{C},\gamma_y)$ for a.e. y, so for some $l \in \mathbf{N}$, $(H * h_i)_{i=1}^l$ is an ϵ-net for $H * B_y$ for every y outside of a set $D \in \mathcal{B}$ with $\nu(D) < \epsilon$. Put $g_i = H * h_i$, $1 \le i \le l$.

□

Here now is the second of three theorems from this section to be used later.

Theorem 3.11 If $(Z,\mathcal{C},\gamma,\Omega)$ is a compact extension of $(Y,\mathcal{B},\nu,\Omega)$ with respect to G along $\mathcal{F}^{(1)}$, then there exists a refinement $\mathcal{F}^{(2)} \subset \mathcal{F}^{(1)}$ such that $AP(Z,Y,G,\mathcal{F}^{(2)})$ is dense in $\mathrm{L}^2(Z,\mathcal{C},\gamma)$.

Proof. According to Definition 3.6 and Remark 3.7 (i) there exists a $\frac{Z}{Y}$-kernel H with

$$\text{IP-}\lim_{\alpha\in\mathcal{F}^{(1)}} T(\alpha)H = H$$

for all $T \in G$, and

$$\mathrm{L}^2(Z,\mathcal{C},\gamma) = \overline{\{H * \phi :\ \phi \in \mathrm{L}^2(Z,\mathcal{C},\gamma)\}},$$

By Proposition 3.8, we may select a refinement $\mathcal{F}^{(2)} \subset \mathcal{F}^{(1)}$ having the property that for every $T(\alpha_1,\cdots,\alpha_m) \in \mathbf{PE}(G,\mathcal{F}^{(2)})$ and every $\epsilon > 0$ there exists $\alpha_0 \in \mathcal{F}^{(2)}$ such that whenever $(\alpha_1,\cdots,\alpha_m) \in (\mathcal{F}^{(2)})^m_<$ with $\alpha_1 > \alpha_0$, we have

$$\big|\big|T(\alpha_1,\cdots,\alpha_m)H - H\big|\big| \le \epsilon.$$

Suppose now that $\psi \in L^\infty(Z,\mathcal{C},\gamma)$. We claim that $H * \psi \in AP(Z,Y,G,\mathcal{F}^{(2)})$. Having established that, the proof will be complete. We may assume that $||\psi||_\infty = 1$.

Let $\epsilon > 0$. By Lemma 3.10 there exist functions $g_1,\cdots,g_l \in \mathrm{L}^2(Z,\mathcal{C},\gamma)$ and a set $D \in \mathcal{B}$ with $\nu(D) < \epsilon$ such that for all f with $\big|\big|f\big|\big|_\infty < 1$ and all $y \in D^c$ there exists $j = j(f,y)$ such that

$$\big|\big|H * f - g_j\big|\big|_y < \frac{\epsilon}{2}.$$

Suppose now that $\delta > 0$ and $T(\alpha_1,\cdots,\alpha_m) \in \mathbf{PE}(G,\mathcal{F}^{(2)})$. There exists $\alpha_0 \in \mathcal{F}^{(2)}$ such that whenever $(\alpha_1,\cdots,\alpha_m) \in (\mathcal{F}^{(2)})^m_<$, $\alpha_1 > \alpha_0$, we have

$$\big|\big|T(\alpha_1,\cdots,\alpha_m)H - H\big|\big| < \frac{\delta\epsilon^2}{4}.$$

This implies that there exist sets $E(\alpha_1,\cdots,\alpha_m) \in \mathcal{B}$ of measure $< \delta$ having the property that

$$\big|\big|T(\alpha_1,\cdots,\alpha_m)H - H\big|\big|_y < \frac{\epsilon}{2}$$

for all $(\alpha_1,\cdots,\alpha_m) \in \big(\mathcal{F}^{(2)}\big)^m_<$ with $\alpha_1 > \alpha_0$ and $y \in E(\alpha_1,\cdots,\alpha_m)^c$.

Note that $T(\alpha_1,\cdots,\alpha_m)(H * \psi) = \big(T(\alpha_1,\cdots,\alpha_m)H\big) * \big(T(\alpha_1,\cdots,\alpha_m)\psi\big)$. Therefore, if $y \in \big(D \cup E(\alpha_1,\cdots,\alpha_m)\big)^c$ then

$$\begin{aligned}&\big|\big|T(\alpha_1,\cdots,\alpha_m)(H * \psi) - g_{j(T(\alpha_1,\cdots,\alpha_m)\psi,y)}\big|\big|_y\\ \le&\big|\big|\big(T(\alpha_1,\cdots,\alpha_m)H\big) * \big(T(\alpha_1,\cdots,\alpha_m)\psi\big) - H * \big(T(\alpha_1,\cdots,\alpha_m)\psi\big)\big|\big|_y\\ &+ \big|\big|H * T(\alpha_1,\cdots,\alpha_m)\psi - g_{j(T(\alpha_1,\cdots,\alpha_m)\psi,y)}\big|\big|_y < \frac{\epsilon}{2} + \frac{\epsilon}{2} = \epsilon.\end{aligned}$$

It follows that $H * \psi \in AP(Z,Y,G,\mathcal{F}^{(2)})$.

□

Having established the density of almost periodic functions for primitive extensions, we turn to the matter of establishing their existence. The following lemma illustrates how one may replace an arbitrary kernel by a self-adjoint non-negative definite one.

Lemma 3.12 Let $(Z, \mathcal{C}, \gamma, \Omega)$ be an extension of $(Y, \mathcal{B}, \nu, \Omega)$. Suppose that $0 \not\equiv H \in \mathrm{L}^2(Z \times_Y Z, \mathcal{C} \otimes_{\mathcal{B}} \mathcal{C}, \tilde{\gamma})$ satisfies $E(H|Y) = 0$. Let $d \in \mathbf{N}$ and let $T \in \mathbf{PE}_{\leq d}(1)$. If $\mathcal{F}^{(1)}$ is an IP-ring with the property that

$$\operatorname*{IP-lim}_{\alpha \in \mathcal{F}^{(1)}} T(\alpha) H = H,$$

then there exists a non-trivial self-adjoint non-negative definite $\frac{Z}{Y}$-kernel K with

$$\operatorname*{IP-lim}_{\alpha \in \mathcal{F}^{(1)}} T(\alpha) K = K.$$

Proof. Truncating the function $H(z_1, z_2)$ if necessary, we may assume without loss of generality that H is bounded. Also we may assume that $H(z_1, z_2)$ is not a function of z_2 alone (otherwise, since H is not constant, H is not a function of z_1 alone and proceed similarly). In this case,

$$\tilde{H}(z_1, z_2) = H(z_1, z_2) - \int H(t, z_2)\, d\gamma_y(t)$$

is not a function of z_2 alone. Furthermore, for fixed z_2,

$$\int \tilde{H}(z_1, z_2)\, d\gamma_y(z_1) = 0,$$

so that $\tilde{H}$ is a non-trivial $\frac{Z}{Y}$-kernel which satisfies, as is easily checked,

$$\operatorname*{IP-lim}_{\alpha \in \mathcal{F}^{(1)}} T(\alpha) \tilde{H} = \tilde{H}.$$

For z_1, z_2 in the fiber over y we let

$$K(z_1, z_2) = \int \tilde{H}(z_1, t) \overline{\tilde{H}(z_2, t)}\, d\gamma_y(t).$$

$K(z_1, z_2)$ is self-adjoint and $K = \tilde{H}\tilde{H}^*$ as an operator, hence it is non-negative definite. K is a $\frac{Z}{Y}$-kernel, for

$$\begin{aligned}\int K(z_1, z_2)\, d\gamma_y(z_1) &= \int \int \tilde{H}(z_1, t) \overline{\tilde{H}(z_2, t)}\, d\gamma_y(t)\, d\gamma_y(z_1) \\ &= \int \overline{\tilde{H}(z_2, t)} \int \tilde{H}(z_1, t)\, d\gamma_y(z_1)\, d\gamma_y(t) = 0.\end{aligned}$$

We claim that

$$\operatorname*{IP-lim}_{\alpha \in \mathcal{F}^{(1)}} T(\alpha) K = K.$$

To see this, note that

$$\int \int \int \left| T(\alpha) \tilde{H}(z, t) - \tilde{H}(z, t) \right|^2 d\gamma_y(t)\, d\gamma_y(z) d\nu(y) \to 0.$$

Therefore, for all α sufficiently far out, most y's (with respect to ν) will have the property that for most z's (with respect to γ_y,)

$$\left|\left| T(\alpha) \tilde{H}(z, \cdot) - \tilde{H}(z, \cdot) \right|\right|_y^2 = \int \left| T(\alpha) \tilde{H}(z, t) - \tilde{H}(z, t) \right|^2 d\gamma_y(t)$$

is small, which implies that for most pairs (z_1, z_2) (with respect to $\gamma_y \times \gamma_y$,)

$$\left|T(\alpha)K(z_1,z_2)-K(z_1,z_2)\right| = \left|\langle T(\alpha)\tilde{H}(z_1,\cdot),T(\alpha)\tilde{H}(z_2,\cdot)\rangle_y - \langle \tilde{H}(z_1,\cdot),\tilde{H}(z_2,\cdot)\rangle_y\right|$$

is small, which is what we need.

□

The next two lemmas demonstrate the impossibility of trivializing a non-negative definite, self-adjoint kernel by passing to an IP-limit along a compact polynomial expression.

Lemma 3.13 Suppose that $(Z, \mathcal{C}, \gamma, \Omega)$ is a compact extension of $(Y, \mathcal{B}, \nu, \Omega)$ with respect to G along $\mathcal{F}^{(1)}$. Suppose that $T \in G$ and $f \in \mathrm{L}^2(Z, \mathcal{C}, \gamma)$ satisfy

$$\operatorname*{IP-lim}_{\alpha\in\mathcal{F}^{(1)}} \int \big(T(\alpha)f \otimes T(\alpha)\overline{f}\big)(g \otimes \overline{g})\, d\tilde{\gamma} = 0$$

for all $g \in L^\infty(Z, \mathcal{C}, \gamma)$. Then $f = 0$.

Proof. Let δ with $0 < \delta < 1$ be small enough that $||f - h||^2 < 2\delta$ implies $|\langle f, h\rangle| > \frac{||f||^2}{2}$. Let $\epsilon > 0$ be small enough that $\epsilon^2 < \delta$ and

$$\int_E ||f||_y^2 d\nu(y) < \delta \text{ for all } E \in \mathcal{B} \text{ with } \nu(E) < \epsilon. \tag{3.2}$$

Choose by Theorem 3.11 a refinement $\mathcal{F}^{(2)} \subset \mathcal{F}^{(1)}$ such that $AP(Z, Y, G, \mathcal{F}^{(2)})$ is dense in $\mathrm{L}^2(Z, \mathcal{C}, \gamma)$. Since $f \in \overline{AP(Z, Y, G, \mathcal{F}^{(2)})}$ we may by Remark 3.2 (iv) find $g_1, \cdots, g_l \in L^\infty(Z, \mathcal{C}, \gamma)$ and $\alpha_0 \in \mathcal{F}^{(2)}$ such that whenever $\alpha > \alpha_0$ we have a set $E(\alpha) \in \mathcal{B}$, $\nu\big(E(\alpha)\big) < \epsilon$, with the property that for all $y \in E(\alpha)^c$, there exists $j = j(\alpha, y)$ with $1 \leq j \leq l$ such that

$$||T(\alpha)f - g_{j(\alpha,y)}||_y < \epsilon,\ y \in E(\alpha)^c. \tag{3.3}$$

For every $\alpha \in \mathcal{F}^{(2)}$, $\alpha > \alpha_0$, let h_α be the function equal to $g_{j(\alpha,y)}$ on the fiber over y when $y \in E(\alpha)^c$, and equal to zero on fibers over $y \in E(\alpha)$. Each h_α is measurable, and by (3.2) and (3.3) we have

$$\begin{aligned} ||T(\alpha)f - h_\alpha||^2 &= \int\int |T(\alpha)f - h_\alpha|^2\, d\gamma_y\, d\nu(y) \\ &= \int_{E(\alpha)} ||T(\alpha)f||_y^2\, d\nu(y) + \int_{E(\alpha)^c} ||T(\alpha)f - g_{j(\alpha,y)}||_y^2\, d\nu(y) \\ &\leq \delta + \epsilon^2 < 2\delta. \end{aligned}$$

Therefore we also have $||f - T(\alpha)^{-1}h_\alpha||^2 < 2\delta$, whence

$$\left|\langle T(\alpha)f, h_\alpha\rangle\right| = \left|\langle f, T(\alpha)^{-1}h_\alpha\rangle\right| \geq \frac{||f||^2}{2}. \tag{3.4}$$

On the other hand,

$$\begin{aligned}
\left|\langle T(\alpha)f, h_\alpha\rangle\right| &= \left|\int\int T(\alpha)f\overline{h}_\alpha\, d\gamma_y\, d\nu(y)\right| \\
&\leq \int\left|\int T(\alpha)f\overline{h}_\alpha\, d\gamma_y\right| d\nu(y) \\
&\leq \int\sum_{j=1}^{l}\left|\int T(\alpha)f\overline{g}_j\, d\gamma_y\right| d\nu(y) \\
&\leq \sum_{j=1}^{l}\left(\int\left|\int T(\alpha)f\overline{g}_j\, d\gamma_y\right|^2 d\nu(y)\right)^{\frac{1}{2}} \\
&= \sum_{j=1}^{l}\left(\int\big(T(\alpha)f\otimes T(\alpha)\overline{f}\big)(\overline{g}\otimes g)\, d\tilde{\gamma}\right)^{\frac{1}{2}} \to 0.
\end{aligned}$$

It follows that $||f|| = 0$, as desired.

□

Lemma 3.14, which is what we really need, is just an elaboration on Lemma 3.13.

Lemma 3.14 Suppose that $(Z, \mathcal{C}, \gamma, \Omega)$ is a compact extension of $(Y, \mathcal{B}, \nu, \Omega)$ with respect to G along $\mathcal{F}^{(1)}$. If $T \in G$ and H is a self-adjoint, non-negative definite $\frac{Z}{Y}$-kernel satisfying

$$\operatorname*{IP-lim}_{\alpha\in\mathcal{F}^{(1)}} \int \big(T(\alpha)H\big)(g\otimes\overline{g})\, d\tilde{\gamma} = 0 \tag{3.5}$$

for all $g \in L^\infty(Z, \mathcal{C}, \gamma)$ then $H \equiv 0$.

Proof. Suppose that H is given by

$$H(z_1, z_2) = \sum_k \lambda_k(y)\phi_k(z_1)\overline{\phi_k(z_2)}.$$

For any $j \in \mathbf{N}$ we have

$$\begin{aligned}
&\int \big(T(\alpha)H\big)(g\otimes\overline{g})\, d\tilde{\gamma} \\
&= \sum_k \int \lambda_k(y)\big(T(\alpha)\phi_k\otimes T(\alpha)\overline{\phi_k}\big)(g\otimes\overline{g})\, d\tilde{\gamma} \\
&= \sum_k \int \lambda_k(y)\left|\int T(\alpha)\phi_k\, g\, d\gamma_y\right|^2 d\nu(y) \qquad (3.6)\\
&\geq \int \lambda_j(y)\left|\int T(\alpha)\phi_j\, g\, d\gamma_y\right|^2 d\nu(y) \\
&= \int \big(T(\alpha)(\lambda_j^{\frac{1}{2}}\phi_j)\otimes T(\alpha)(\lambda_j^{\frac{1}{2}}\overline{\phi_j})\big)(g\otimes\overline{g})\, d\tilde{\gamma}.
\end{aligned}$$

In light of (3.5) and (3.6) we get that

$$\operatorname*{IP-lim}_{\alpha\in\mathcal{F}^{(1)}} \int \big(T(\alpha)(\lambda_j^{\frac{1}{2}}\phi_j)\otimes T(\alpha)(\lambda_j^{\frac{1}{2}}\overline{\phi_j})\big)(g\otimes\overline{g})\, d\tilde{\gamma} = 0,$$

which together with Lemma 3.13, gives $\lambda_j^{\frac{1}{2}}\phi_j = 0$ for all j, which implies that

$$H(z_1, z_2) = \sum_k \lambda_k(y)\phi_k(z_1)\overline{\phi_k(z_2)} = 0.$$

□

Finally the groundwork has been laid for establishing the existence of primitive extensions. Here is the idea: suppose that a system $(Y, \mathcal{B}, \nu, \Omega)$ is a factor of another system $(X, \mathcal{A}, \mu, \Omega)$ and suppose that $\mathcal{F}^{(1)}$ is an IP-ring. Suppose $d \in \mathbf{N}$ and let $G < \mathbf{PE}_{\leq d}(1)$ be a subgroup maximal with respect to the property that for some refinement $\mathcal{F}^{(2)} \subset \mathcal{F}^{(1)}$ and some non-trivial self-adjoint non-negative definite $\frac{X}{Y}$-kernel H we have, for all $T \in G$,

$$\operatorname*{IP-lim}_{\alpha \in \mathcal{F}^{(2)}} T(\alpha)H = H.$$

Let $(Z, \mathcal{C}, \gamma, \Omega)$ be the resulting extension of $(Y, \mathcal{B}, \nu, \Omega)$ which is compact with respect to G along $\mathcal{F}^{(2)}$. By Proposition 2.14 there exists a refinement $\mathcal{F}^{(3)} \subset \mathcal{F}^{(2)}$ such that for every $B \in \mathrm{L}^2(Z \times_Y Z, \mathcal{C} \otimes_{\mathcal{B}} \mathcal{C}, \tilde{\gamma})$ and every $T \in \mathbf{PE}_{\leq d}(1)$

$$P_T B = \operatorname*{IP-lim}_{\alpha \in \mathcal{F}^{(3)}} T(\alpha)B$$

exists weakly (since $\mathbf{PE}_{\leq d}(1)$ is countable and $\mathrm{L}^2(Z \times_Y Z, \mathcal{C} \otimes_{\mathcal{B}} \mathcal{C}, \tilde{\gamma})$ is separable). By Theorem 2.17, the operators $\{P_T : T \in \mathbf{PE}_{\leq d}(1)\}$ are commuting orthogonal projections.

Let $T \in \big(\mathbf{PE}_{\leq d}(1) \setminus G\big)$. We claim that T is mixing on Z relative to Y along $\mathcal{F}^{(3)}$. To prove this, we must show that for arbitrary $B \in \mathrm{L}^2(Z \times_Y Z, \mathcal{C} \otimes_{\mathcal{B}} \mathcal{C}, \tilde{\gamma})$ satisfying $E(B|Y) = 0$,

$$P_T B = \operatorname*{IP-lim}_{\alpha \in \mathcal{F}^{(3)}} T(\alpha)B = 0.$$

Suppose then that for some such B, $P_T B \neq 0$. By idempotence of P_T, we have

$$\operatorname*{IP-lim}_{\alpha \in \mathcal{F}^{(3)}} T(\alpha)P_T B = P_T(P_T B) = P_T B,$$

so by Lemma 3.12 there exists a non-trivial, self-adjoint, non-negative definite $\frac{Z}{Y}$-kernel K with

$$P_T K = \operatorname*{IP-lim}_{\alpha \in \mathcal{F}^{(3)}} T(\alpha)K = K.$$

Let $\{S_1, \cdots, S_j\}$ be a generating set for G. One easily checks that the property of being a self-adjoint non-negative definite $\frac{Z}{Y}$-kernel is preserved under the action of Ω and under passage to weak limits, so

$$L = P_{S_1} \cdots P_{S_j} K$$

is a self-adjoint, non-negative definite $\frac{Z}{Y}$-kernel. Meanwhile by Lemma 3.14 L is non-trivial, and for every S in the group generated by $\{T, S_1, \cdots, S_j\}$, $P_S L = L$. This contradicts the maximality of G. We have thus proved (writing $\mathcal{F}^{(2)}$ for $\mathcal{F}^{(3)}$) the following theorem (which is the third of three theorems from this section to be used in Section 5).

Theorem 3.15 Let $d \in \mathbf{N}$. If $(Y, \mathcal{B}, \nu, \Omega)$ is a proper factor of $(X, \mathcal{A}, \mu, \Omega)$ and $\mathcal{F}^{(1)}$ is any IP-ring then there exist a factor $(Z, \mathcal{C}, \gamma, \Omega)$ of $(X, \mathcal{A}, \mu, \Omega)$, a subgroup $G < \mathbf{PE}_{\leq d}(1)$, and a refinement $\mathcal{F}^{(2)} \subset \mathcal{F}^{(1)}$ such that $(Z, \mathcal{C}, \gamma, \Omega)$ is a primitive extension of $(Y, \mathcal{B}, \nu, \Omega)$ along $\mathcal{F}^{(2)}$ with compact part G.

CHAPTER 4

RELATIVE POLYNOMIAL MIXING

We now will treat the role of mixing in primitive extensions. The theorem we are at the moment aiming for is Theorem 4.10. A brief sketch of our plan is as follows: in Proposition 4.1 we give the most fundamental consequence of relative mixing, and Proposition 4.2 is a natural extension of this. After a few examples serving to illustrate the inductive setup we will use in the proof of Theorem 4.10, we introduce a notion which generalizes that of polynomial expression and prove a couple of lemmas (Lemma 4.8 and Lemma 4.9) centered around this notion. A large part of these efforts are undertaken in order to manage the algebraical nature of our present dealings, namely with regard to a primitive extension whose compact part is a subgroup G of $\mathbf{PE}_{\leq d}(1)$. Finally we prove Theorem 4.10 and its corollary, Theorem 4.12, which is what we will need for our proof of Theorem 1.3.

Proposition 4.1 If $\mathcal{F}^{(1)}$ is an IP-ring, $(Y,\mathcal{B},\nu,\Omega)$ is a factor of $(Z,\mathcal{C},\gamma,\Omega)$, and $T\in\mathbf{PE}_{\leq d}(1)$ is mixing on Z relative to Y along $\mathcal{F}^{(1)}$, then if $f,g\in \mathrm{L}^2(Z,\mathcal{C},\gamma)$ with either $E(f|Y)=0$ or $E(g|Y)=0$, then

$$\underset{\beta\in\mathcal{F}^{(1)}}{\text{IP-lim}}\ \big|\big|E(fT(\beta)g|Y)\big|\big| = 0.$$

Proof. We use the fact that T is mixing on $Z\times_Y Z$ relative to Y along $\mathcal{F}^{(1)}$ (see Remark 3.7 (iii)). We have

$$\begin{aligned}
&\underset{\beta\in\mathcal{F}^{(1)}}{\text{IP-lim}}\ \big|\big|E(fT(\beta)g|Y)\big|\big|^2\\
&=\underset{\beta\in\mathcal{F}^{(1)}}{\text{IP-lim}}\ \int\Big|\int fT(\beta)g\,d\gamma_y\Big|^2\,d\nu(y)\\
&=\underset{\beta\in\mathcal{F}^{(1)}}{\text{IP-lim}}\ \int(f\otimes\overline{f})T(\beta)(g\otimes\overline{g})\,d\tilde{\gamma}\\
&=\underset{\beta\in\mathcal{F}^{(1)}}{\text{IP-lim}}\ \int E(f\otimes\overline{f}|Y)T(\beta)E(g\otimes\overline{g}|Y)\,d\nu=0.
\end{aligned}$$

□

Proposition 4.1 shows that if T is a relatively mixing polynomial expression then $T(\beta)g$ approaches fiberwise orthogonality with any f. What we will need is the somewhat stronger fact that $T(\beta)g$ actually approaches fiberwise orthogonality with the whole "compact orbit" of f. This is the content of Proposition 4.2.

Proposition 4.2 If $(Z,\mathcal{C},\gamma,\Omega)$ is a primitive extension of $(Y,\mathcal{B},\nu,\Omega)$ along $\mathcal{F}^{(1)}$ with compact part G, $T\in\big(\mathbf{PE}_{\leq d}(1)\setminus G\big)$, $U^{(\beta)}\in G$ for all $\beta\in\mathcal{F}^{(1)}$, and

$f, g \in L^\infty(Z, \mathcal{C}, \gamma)$ with either $E(f|Y) = 0$ or $E(g|Y) = 0$, then for some IP-ring $\mathcal{F}^{(2)} \subset \mathcal{F}^{(1)}$,

$$\text{IP-}\lim_{(\beta,\alpha)\in(\mathcal{F}^{(2)})^2_<} \Big|\Big|E\big(U^{(\beta)}(\alpha)f\, T(\beta)g|Y\big)\Big|\Big| = 0.$$

Proof. Fix $f, g \in L^\infty(Z, \mathcal{C}, \gamma)$. Using Theorems 2.16 and 3.11, let $\mathcal{F}^{(2)} \subset \mathcal{F}^{(1)}$ be a subring with the property that

$$\text{IP-}\lim_{(\beta,\alpha)\in(\mathcal{F}^{(2)})^2_<} \Big|\Big|E\big(U^{(\beta)}(\alpha)f\, T(\beta)g|Y\big)\Big|\Big|$$

exists and such that $AP(Z, Y, G, \mathcal{F}^{(2)})$ is dense in $L^2(Z, \mathcal{C}, \gamma)$. Let $\epsilon > 0$ and $\alpha_0 \in \mathcal{F}^{(2)}$ be arbitrary. We will find $(\beta, \alpha) \in (\mathcal{F}^{(2)})^2_<$ with $\beta > \alpha_0$ such that $\big|\big|E\big(U^{(\beta)}(\alpha)f\, T(\beta)g|Y\big)\big|\big| \leq \epsilon$. This will suffice for the proof.

Let $\delta > 0$ be so small that

$$\big(\delta^2||g||^2_\infty + 4\delta||f||^2_\infty||g||^2_\infty\big)^{\frac{1}{2}} + \delta < \epsilon.$$

By Remark 3.2 (iv), there exist functions $g_1, \cdots, g_l \in L^\infty(Z, \mathcal{C}, \gamma)$ and an $\mathcal{F}$-sequence $(\alpha_\beta)_{\beta\in\mathcal{F}^{(2)}} \subset \mathcal{F}^{(2)}$ such that for every $\alpha, \beta \in \mathcal{F}^{(2)}$ with $\alpha > \alpha_\beta$ there exists a set $E = E(\alpha, \beta) \in \mathcal{B}$ with $\nu(E) < \delta$ having the property that for each $y \in E^c$ one has a number $j = j(\alpha, \beta, y)$, $1 \leq j \leq l$, such that

$$\big|\big|U^{(\beta)}(\alpha)f - g_j\big|\big|_y < \delta.$$

Moreover, if $E(f|Y) = 0$ a.e. the g_i may be chosen with $E(g_i|Y) = 0$ a.e. (One simply replaces g_i by $g_i - E(g_i|Y)$ and checks that the distance between g_i and $U^{(\beta)}f$ cannot thereby increase in $L^2(Z, \gamma_y)$ for a.e. y.)

Having made this observation, it is now the case that either $E(g|Y) = 0$ or $E(g_i|Y) = 0$ for $i = 1, \cdots, l$. By Proposition 4.1, there exists $\beta_0 \in \mathcal{F}^{(2)}$ with $\beta_0 > \alpha_0$ and having the property that for all $\beta \in \mathcal{F}^{(2)}$ with $\beta > \beta_0$ we have

$$\Big(\sum_{j=1}^{l} ||E(g_j T(\beta)g|Y)||^2\Big)^{\frac{1}{2}} < \delta.$$

Letting $\beta > \beta_0$ and $\alpha > \alpha_\beta$,

$$\begin{aligned}
&\Big|\Big|E\big(U^{(\beta)}(\alpha)f\, T(\beta)g|Y\big)\Big|\Big| \\
\leq&\Big|\Big|E\big((U^{(\beta)}(\alpha)f - g_{j(\alpha,\beta,y)})T(\beta)g|Y\big)\Big|\Big| + \Big|\Big|E\big(g_{j(\alpha,\beta,y)}\, T(\beta)g|Y\big)\Big|\Big| \\
=&\Big(\int\Big|\int (U^{(\beta)}(\alpha)f - g_{j(\alpha,\beta,y)})T(\beta)g\, d\gamma_y\Big|^2 d\nu(y)\Big)^{\frac{1}{2}} \\
&+ \Big(\int\Big|\int g_{j(\alpha,\beta,y)}T(\beta)g\, d\gamma_y\Big|^2 d\nu(y)\Big)^{\frac{1}{2}} \\
\leq&\Big(\int ||U^{(\beta)}(\alpha)f - g_{j(\alpha\beta,y)}||^2_y\, ||g||^2_\infty\, d\nu(y)\Big)^{\frac{1}{2}} \\
&+ \Big(\int\sum_{j=1}^{l}\Big|\int g_j T(\beta)g\, d\gamma_y\Big|^2 d\nu(y)\Big)^{\frac{1}{2}} \\
\leq&\big(\delta^2||g||^2_\infty + 4\delta||f||^2_\infty||g||^2_\infty\big)^{\frac{1}{2}} + \Big(\sum_{j=1}^{l}||E(g_j T(\beta)g|Y)||^2\Big)^{\frac{1}{2}} < \epsilon.
\end{aligned}$$

□

We now introduce the inductive scheme we will be using to establish relative mixing properties, which we call *PET-induction* (for *polynomial exhaustion technique*). A version of such an inductive technique seems to be unavoidable when one deals with multiple recurrence along polynomials. (See, for example, [B1], [BL1], [BL2], [BM].)

Let $d \in \mathbf{N}$. For non-trivial

$$T(\alpha) = \prod_{i=1}^{r} T_i^{p_i(n_\alpha^{(1)},\cdots,n_\alpha^{(k)})} \in \mathbf{PE}_{\leq d}(1), \tag{4.1}$$

we define the *weight* $w(T)$ of T to be the pair (a,b), where $0 \neq b = \deg p_a$ and $p_i = 0$ if $a < i \leq r$ (mneumonic device: "(last T, last degree)"). We write $(a_1,b_1) > (a_2,b_2)$ if $a_1 > a_2$ or if $a_1 = a_2$, $b_1 > b_2$. In the following examples we are using the symbol x_i in place of $n_\alpha^{(i)}$, $1 \leq i \leq k$.

Example 4.3 $T_1^{x_1x_2}T_2^{x_2x_3^2}$ has weight $(2,3)$, and $T_2^{x_2^4}T_5^{4x_3+2x_1x_2}$ has weight $(5,2) > (2,3)$.

If T is given by (4.1) and

$$S(\alpha) = \prod_{i=1}^{r} T_i^{q_i(n_\alpha^{(1)},\cdots,n_\alpha^{(k)})} \in \mathbf{PE}_{\leq d}(1)$$

is another non-trivial polynomial expression, we write $T \sim S$ if T and S have the same weight (a,b), and if furthermore the polynomials p_a and q_a coincide in their b^{th}-degree terms. $\sim$ is an equivalence relation.

Example 4.4 $T_1^{x_1}T_2^{x_1}T_3^{x_1x_2+x_3^2+2x_1} \sim T_2^{x_3}T_3^{x_1x_2+x_3^2+7x_2} \not\sim T_3^{x_1x_2+7x_2}$.

Suppose now that $A = \{S_1,\cdots,S_t\} \subset \mathbf{PE}_{\leq d}(1)$. For each weight (a,b), $1 \leq a \leq r$, $1 \leq b \leq d$, let m_{ab} be the number of equivalence classes (under $\sim$) represented by elements in S of weight (a,b). The $r \times d$ matrix (m_{ab}) will be called the *weight matrix* of A.

Example 4.5 Let $d = 4$ and suppose for the moment that $r = 5$. The weight matrix of the family

$$\{T_1^{x_1x_2}T_2^{x_2x_3^2}, T_2^{x_2^4}T_5^{4x_3+2x_1x_2}, T_1^{x_1}T_2^{x_1}T_3^{x_1x_2+x_3^2+2x_1}, T_2^{x_3}T_3^{x_1x_2+x_3^2+7x_2}, T_3^{x_1x_2+7x_2}\}$$

is

$$\begin{pmatrix} 0 & 0 & 0 & 0 \\ 0 & 0 & 1 & 0 \\ 0 & 2 & 0 & 0 \\ 0 & 0 & 0 & 0 \\ 0 & 1 & 0 & 0 \end{pmatrix}$$

The 2 appearing in the $(3,2)$ position of the above matrix comes from the last three polynomial expressions, which are all of weight (3,2) and which fall into 2 equivalence classes (see Example 4.4 above).

If a finite family $A \subset \mathbf{PE}_{\leq d}(1)$ has weight matrix (m_{ab}) and another finite family $B \subset \mathbf{PE}_{\leq d}(1)$ has weight matrix (n_{ab}), then we will write $A < B$, and say that A *precedes* B, if there exists a weight (i,j) such that $m_{ij} < n_{ij}$ and $m_{ab} = n_{ab}$

whenever $(a,b) > (i,j)$. $<$ is a partial order on the collection of finite families in $\mathbf{PE}_{\leq d}(1)$ which arises from a well-ordering on the set of weight matrices. Therefore, in order to show that some assertion $\mathcal{W}(A)$ holds for all finite sets of polynomial expressions A, it is sufficient to show first that $\mathcal{W}(A)$ holds for any A with the minimal weight matrix

$$\begin{pmatrix} 1 & 0 & \cdots & 0 \\ 0 & 0 & \cdots & 0 \\ \vdots & \vdots & \ddots & \vdots \\ 0 & 0 & \cdots & 0 \end{pmatrix}, \tag{4.2}$$

and then that $\mathcal{W}(A)$ holds provided $\mathcal{W}(B)$ holds for every B preceding A. After some brief preliminaries, we will use this method to prove Theorem 4.10.

Definition 4.6 Suppose $\mathcal{F}^{(1)}$ is an IP-ring, $t \in \mathbf{N}$, $\{p_i(x_1, \cdots, x_k) : 1 \leq i \leq t\}$ is a set of polynomials in $\mathbf{Z}[x_1, \cdots, x_k]$ and $\{y_1, \cdots, y_t\}$ is a basis for a group Γ which is isomorphic to $\mathbf{Z}^t$. A map $\mathcal{F}^{(1)} \to \Gamma$,

$$S^{(\beta)} = \sum_{i=1}^{t} p_i(n_\beta^{(1)}, \cdots, n_\beta^{(k)}) y_i,$$

will be called a *polynomial form of rank* 1 *on* Γ. Suppose that $m \in \mathbf{N}$, and

$$\{p_i\big((x_j^{(b)})_{1 \leq b \leq k,\ 1 \leq j \leq m}\big) : 1 \leq i \leq t\}$$

is a set of polynomials in the $k \times m$-matrix of variables $(x_j^{(b)})_{1 \leq b \leq k, 1 \leq j \leq m}$. A map $(\mathcal{F}^{(1)})^m_< \to \mathbf{Z}^t$,

$$S^{(\beta_1, \cdots, \beta_m)} = \sum_{i=1}^{t} p_i\big((n_{\beta_c}^{(b)})_{1 \leq b \leq k,\ 1 \leq c \leq m}\big) y_i,$$

will be called a *polynomial form of rank* m *on* Γ. In either case, if the polynomials p_i have zero constant term, the map will be called an *integral polynomial form.*

Note that the "polynomial expressions" $T(\alpha) = \prod_{i=1}^r T_i^{p_i(n_\alpha^{(1)}, \cdots, n_\alpha^{(k)})}$ may themselves be seen as polynomial forms on $\mathbf{Z}^r$. One reason for making a distinction between the two notions is the different role we have in mind for "polynomial forms" (which will always for us have range in $\mathbf{PE}_{\leq d}(1)$, in contrast to polynomial expressions, which have range lying in Ω).

Remarks 4.7

(i) Any polynomial form expressed in terms of one basis $\{y_1, \cdots, y_t\}$ may be expressed in the same manner as above in terms of any other basis (with different polynomials). Since $\mathbf{PE}_{\leq d}(1)$ is isomorphic to $\mathbf{Z}^t$ (for some t), we may speak of polynomial forms of rank m on $\mathbf{PE}_{\leq d}(1)$, or on a subgroup $G < \mathbf{PE}_{\leq d}(1)$. It is easily seen that any such polynomial form $S^{(\beta_1, \cdots, \beta_m)}$ must be of the form $S^{(\beta_1, \cdots, \beta_m)}(\alpha) = T(\beta_1, \cdots, \beta_m, \alpha)$, where $T \in \mathbf{PE}(m+1)$. Conversely, if $T \in \mathbf{PE}_{\leq d}(m+1)$ and we define $S^{(\beta_1, \cdots, \beta_m)}(\alpha) = T(\beta_1, \cdots, \beta_m, \alpha)$ then $S^{(\beta_1, \cdots, \beta_m)}$ will be a polynomial form on $\mathbf{PE}_{\leq d}(1)$.

(ii) If $S^{(\beta_1, \cdots, \beta_t)}$ is an integral polynomial form on Γ and H is a subgroup of Γ of finite index, $[\Gamma : H] < \infty$, and $\mathcal{F}^{(1)}$ is any IP-ring, then restricted to some suitable refinement $\mathcal{F}^{(2)} \subset \mathcal{F}^{(1)}$, $S^{(\beta_1, \cdots, \beta_t)}$ is an integral polynomial form on H. (This is another consequence of Theorem 2.13.)

We will now take the opportunity to discuss the polynomial exhaustion technique in a bit more detail. The most fundamental use of the technique (see [B2]) involves a single polynomial $p(n) \in \mathbf{Z}[n]$ and the simple observation that if $r \in \mathbf{Z}$ is fixed then, letting $q(n) = p(n+r) - p(n)$, one has $\deg q < \deg p$. If it is more convenient to work with integral polynomials, i.e. $p(0) = 0$, then letting $q(n) = p(n+r) - p(n) - p(r)$ we have $\deg q < \deg p$ and $q(0) = 0$. For example, if $p(n) = n^3 + n^2$ and $r = 1$ then $q(n) = (n+1)^3 + (n+1)^2 - n^3 - n^2 - 2 = 3n^2 + 5n$.

To extend this idea, suppose $p(n_1, \cdots, n_k) \in \mathbf{Z}[n_1, \cdots, n_k]$ with $p(0, \cdots, 0) = 0$ and let $(r_1, \cdots, r_k) \in \mathbf{Z}^k$. If we put

$$q(n_1, \cdots, n_k) = p(n_1 + r_1, \cdots, n_k + r_k) - p(n_1, \cdots, n_k) - p(r_1, \cdots, r_k),$$

then $q(0, \cdots, 0) = 0$ and $\deg q < \deg p$. (To prove this, consider that it suffices to show it for monomials $p(n_1, \cdots, n_k) = n_1^{a_1} \cdots n_k^{a_k}$, which is easy.)

Let $d \in \mathbf{N}$ and let $S(\alpha) = \prod_{i=1}^r T_i^{p_i(n_\alpha^{(1)}, \cdots, n_\alpha^{(k)})} \in \mathbf{PE}_{\leq d}(1)$. For fixed $\beta \in \mathcal{F}$, set $S^{(\beta)}(\alpha) = S(\alpha \cup \beta) S(\alpha)^{-1} S(\beta)^{-1}$. (Recalling equation (2.6) of Definition 2.5, this is equivalent to saying that $S^{(\beta)}(\alpha) = S^{(2)}(\alpha, \beta)$.) Then $S^{(\beta)}$ is a polynomial form on $\mathbf{PE}_{\leq d}(1)$. Moreover, $S^{(\beta)}(\alpha) = \prod_{i=1}^r T_i^{q_i^{(\beta)}(n_\alpha^{(1)}, \cdots, n_\alpha^{(k)})}$, where $q_i^{(\beta)}(n_\alpha^{(1)}, \cdots, n_\alpha^{(k)}) = p_i(n_\alpha^{(1)} + n_\beta^{(1)}, \cdots, n_\alpha^{(k)} + n_\beta^{(k)}) - p_i(n_\alpha^{(1)}, \cdots, n_\alpha^{(k)}) - p_i(n_\beta^{(1)}, \cdots, n_\beta^{(k)})$. By our earlier observations, namely that $\deg q_i^{(\beta)} < \deg p_i$, $1 \leq i \leq r$, we have $w(S^{(\beta)}) < w(S)$. This decrease in weight lies at the heart of our inductive procedure.

Lemma 4.8 Suppose $t \in \mathbf{N}$ and $G < \mathbf{Z}^t$ is a subgroup. Then there exists a subgroup $J < \mathbf{Z}^t$ such that $J \cap G = \{e\}$ and $[\mathbf{Z}^t : G \oplus J] < \infty$.

Proof. Among all groups $H < \mathbf{Z}^t$ satisfying $H \cap G = \{e\}$, let J be chosen so that the free portion of the quotient group $\frac{\mathbf{Z}^t}{G \oplus J}$ is of minimal dimension. We claim this dimension is zero. Otherwise, there exists some $x \in \mathbf{Z}^t$ such that for all $n \in \mathbf{Z}$, $x^n \notin G \oplus J$. It follows that the group J' generated by J and x satisfies $J' \cap G = \{e\}$, and the free portion of $\frac{\mathbf{Z}^t}{G \oplus J'}$ has dimension one less than the free portion of $\frac{\mathbf{Z}^t}{G \oplus J}$, a contradiction. Therefore $\frac{\mathbf{Z}^t}{G \oplus J}$ is a torsion group, and, being finitely generated, is finite.

□

The conclusion of Lemma 4.8 obviously continues to hold if $\mathbf{Z}^t$ is replaced by any finitely generated free abelian group, in particular a subgroup of $\mathbf{PE}_{\leq d}(1)$.

Lemma 4.9 Suppose $G < \mathbf{PE}_{\leq d}(1)$. There exists a subgroup $J < \mathbf{PE}_{\leq d}(1)$ with $G \cap J = \{I\}$ and $[\mathbf{PE}_{\leq d}(1) : G \oplus J] < \infty$, such that $w(WU) \geq w(W)$ for any $U \in G$ and $W \in J$.

Proof. Let $w_1 < w_2 < \cdots < w_b$ be the complete set of weights occuring in $\mathbf{PE}_{\leq d}(1)$. For $1 \leq i \leq b$, put

$$H_i = \{T \in \mathbf{PE}_{\leq d}(1) : \ w(T) \leq w_i\}.$$

Then $H_1 < H_2 < \cdots < H_b = \mathbf{PE}_{\leq d}(1)$. Also put $G_i = H_i \cap G$, so that $G_1 < \cdots < G_b = G$.

By Lemma 4.8 there exists a subgroup $J_1 < H_1$ such that $G_1 \cap J_1 = \{I\}$ and $[H_1 : (G_1 \oplus J_1)] < \infty$. We have $J_1 \cap G_2 = \{I\}$ and $(G_2 \oplus J_1) < H_2$. Again by Lemma 4.8 there exists $J_2 < H_2$ with $J_2 \cap (G_2 \oplus J_1) = \{I\}$ (so that $G_2 \cap (J_1 \oplus J_2) =$

$\{I\}$) and $[H_2 : (G_2 \oplus J_1) \oplus J_2] < \infty$. We now have $G_3 \cap (J_1 \oplus J_2) = \{I\}$ and $(G_3 \oplus J_1 \oplus J_2) < H_3$. There exists $J_3 < H_3$ with $J_3 \cap (G_3 \oplus J_1 \oplus J_2) = \{I\}$ and $[H_3 : (G_3 \oplus J_1 \oplus J_2) \oplus J_3] < \infty$. Continue in this fashion until $J_1, J_2, \cdots, J_b$ have been chosen. Let $J = J_1 \oplus J_2 \oplus \cdots \oplus J_b$. Then $G \cap J = \{I\}$ and $[\mathbf{PE}_{\leq d}(1) : G \oplus J] < \infty$.

Suppose now we are given $U \in G$ and $W \in J$. Suppose $w(W) = w_i$. Then $W \notin (G \oplus J_1 \oplus J_2 \oplus \cdots \oplus J_{i-1})$, from which it follows that $WU \notin (G \oplus J_1 \oplus J_2 \oplus \cdots \oplus J_{i-1})$. But $WU \in (G \oplus J_1 \oplus J_2 \oplus \cdots \oplus J_i)$. Therefore $w(WU) \geq w_i = w(W)$.

□

We are now ready to use our inductive method to prove that "relative mixing implies polynomial relative mixing of all orders". We give four versions of the statement, not out of intrisic interest (only the second is utilized) but because of the method of proof, which proceeds by a helical string of implications involving these four formulations along an increasing sequence of orders corresponding to the weight matrices of the systems under consideration.

Theorem 4.10 Let $d \in \mathbf{N}$.

(1) If $(Z, \mathcal{C}, \gamma, \Omega)$ is a primitive extension of $(Y, \mathcal{B}, \nu, \Omega)$ along $\mathcal{F}^{(1)}$ with compact part G, $\{f_0, f_1, \cdots, f_t\} \subset L^\infty(Z, \mathcal{C}, \gamma)$, and $\{S_1, \cdots, S_t\} \subset \mathbf{PE}_{\leq d}(1)$ is a mixing set, then setting $S_0 = I$, there exists a refinement $\mathcal{F}^{(2)} \subset \mathcal{F}^{(1)}$ such that

$$\operatorname*{IP-lim}_{\alpha \in \mathcal{F}^{(2)}} \left(\int_Z \prod_{i=0}^{t} S_i(\alpha) f_i \, d\gamma - \int_Y \prod_{i=0}^{t} S_i(\alpha) E(f_i|Y) \, d\nu \right) = 0.$$

(2) If $(Z, \mathcal{C}, \gamma, \Omega)$ is a primitive extension of $(Y, \mathcal{B}, \nu, \Omega)$ along $\mathcal{F}^{(1)}$ with compact part G, $\{f_0, f_1, \cdots, f_t\} \subset L^\infty(Z, \mathcal{C}, \gamma)$, and $\{S_1, \cdots, S_t\} \subset \mathbf{PE}_{\leq d}(1)$ is a mixing set, then setting $S_0 = I$, there exists a refinement $\mathcal{F}^{(2)} \subset \mathcal{F}^{(1)}$ such that

$$\operatorname*{IP-lim}_{\alpha \in \mathcal{F}^{(2)}} \left|\left| E\Big(\prod_{i=0}^{t} S_i(\alpha) f_i | Y \Big) - \prod_{i=0}^{t} S_i(\alpha) E(f_i|Y) \right|\right| = 0.$$

(3) If $(Z, \mathcal{C}, \gamma, \Omega)$ is a primitive extension of $(Y, \mathcal{B}, \nu, \Omega)$ along $\mathcal{F}^{(1)}$ with compact part G, $\{f_i^{(j)} : 0 \leq i \leq t, \ 0 \leq j \leq s\} \subset L^\infty(Z, \mathcal{C}, \gamma)$, $\{U_i^{(j)} : 0 \leq i \leq t, \ 0 \leq j \leq s\} \subset G$, and $\{S_1, \cdots, S_t\} \subset \mathbf{PE}_{\leq d}(1)$ is a mixing set, then setting $S_0 = I$, there exists a refinement $\mathcal{F}^{(2)} \subset \mathcal{F}^{(1)}$ such that

$$\operatorname*{IP-lim}_{\alpha \in \mathcal{F}^{(2)}} \left|\left| E\Big(\prod_{i=0}^{t} S_i(\alpha) \big(\prod_{j=0}^{s} U_i^{(j)}(\alpha) f_i^{(j)} \big) | Y \Big) - \prod_{i=0}^{t} S_i(\alpha) E\Big(\prod_{j=0}^{s} U_i^{(j)}(\alpha) f_i^{(j)} | Y \Big) \right|\right| = 0.$$

(4) If $(Z, \mathcal{C}, \gamma, \Omega)$ is a primitive extension of $(Y, \mathcal{B}, \nu, \Omega)$ along $\mathcal{F}^{(1)}$ with compact part G, $\{f_i^{(j)} : 0 \leq i \leq t, \ 0 \leq j \leq s\} \subset L^\infty(Z, \mathcal{C}, \gamma)$, $\{U_i^{(j)} : 0 \leq i \leq t, \ 0 \leq j \leq s\} \subset G$, and $\{S_1, \cdots, S_t\} \subset \mathbf{PE}_{\leq d}(1)$ is a mixing set then setting $S_0 = I$, there exists a refinement $\mathcal{F}^{(2)} \subset \mathcal{F}^{(1)}$ such that

$$\operatorname*{IP-lim}_{\alpha \in \mathcal{F}^{(2)}} \left(\int_Z \prod_{i=0}^{t} S_i(\alpha) \Big(\prod_{j=0}^{s} U_i^{(j)}(\alpha) f_i^{(j)} \Big) \, d\gamma - \int_Y \prod_{i=0}^{t} S_i(\alpha) E\Big(\prod_{j=0}^{s} U_i^{(j)}(\alpha) f_i^{(j)} | Y \Big) \, d\nu \right) = 0.$$

Proof. The proof is by induction on the weight matrix of the set $A = \{S_1, \cdots, S_t\} \subset \mathbf{PE}_{\leq d}(1)$ using the ordering established earlier in this section. Having established a particular case for a certain weight matrix, validity of that case will be assumed for mixing families of that particular weight matrix in *any* primitive extension $(Z, \mathcal{C}, \gamma, \Omega)$ of $(Y, \mathcal{B}, \nu, \Omega)$. In particular, for mixing families of that weight matrix in the product system $(Z \times_Y Z, \mathcal{C} \otimes \mathcal{C}, \gamma \times_Y \gamma, \Omega)$. (This will only be important in passing from property (1) to property (2).)

First we see that (1) holds when A has minimal weight matrix as given by (4.2). In this case, every polynomial expression in A has the form $S(\alpha) = T_1^{p(n_\alpha^{(1)}, \cdots, n_\alpha^{(k)})}$, with $\deg p = 1$. Furthermore, given two members of A, say $S_1(\alpha) = T_1^{p_1(n_\alpha^{(1)}, \cdots, n_\alpha^{(k)})}$ and $S_2(\alpha) = T_1^{p_2(n_\alpha^{(1)}, \cdots, n_\alpha^{(k)})}$, the polynomials p_1 and p_2 must agree in their first degree terms. Since they are integral polynomials ($p_1(0, \cdots, 0) = p_2(0, \cdots, 0) = 0$), we have $p_1 = p_2$. In other words, A has only one element, which we denote by $S(\alpha) = T_1^{p(n_\alpha^{(1)}, \cdots, n_\alpha^{(k)})}$. By employing the identity

$$\begin{aligned} & f_0 S(\alpha) f_1 - E(f_0|Y) S(\alpha) E(f_1|Y) \\ = & \big(f_0 - E(f_0|Y)\big) S(\alpha) E(f_1|Y) + f_0 S(\alpha) \big(f_1 - E(f_1|Y)\big), \end{aligned}$$

we may suppose that either $E(f_0|Y) \equiv 0$ or $E(f_1|Y) \equiv 0$. Under this assumption it is easy to see that (1) follows from Proposition 4.1, completing the initial case.

Our strategy is now as follows: we first show that (1) holds for all A of a given weight matrix provided (4) holds for every finite set of polynomial expressions B preceding A. Then, for a given weight matrix, we proceed to show that (1) $\rightarrow$ (2) $\rightarrow$ (3) $\rightarrow$ (4) (for A of this weight matrix). When this has been achieved, the proof will have been completed.

(4) $\rightarrow$ (1). We let $A = \{S_1, \cdots, S_t\}$ be a mixing set and assume that (4) is valid for every mixing set which precedes A. Without loss of generality we may assume that $||f_i||_\infty \leq 1$, $0 \leq i \leq t$. Initially, we will also assume that for some i_0, $0 \leq i_0 \leq t$, $E(f_{i_0}|Y) \equiv 0$. Let $\mathcal{F}^{(2)}$ be a refinement of $\mathcal{F}^{(1)}$ such that

$$\operatorname*{IP-lim}_{\alpha \in \mathcal{F}^{(2)}} \int_Z \prod_{i=0}^t S_i(\alpha) f_i \, d\gamma$$

exists. We must show that this limit is zero. It is sufficient to show that for some refinement $\mathcal{F}^{(3)}$ of $\mathcal{F}^{(2)}$,

$$\operatorname*{IP-lim}_{\alpha \in \mathcal{F}^{(3)}} \prod_{i=1}^t S_i(\alpha) f_i = 0$$

weakly. By Proposition 2.18, we can get this by exhibiting a refinement $\mathcal{F}^{(3)} \subset \mathcal{F}^{(2)}$ with (again, see Definition 2.5 for the notation $S_i^{(2)}$)

$$\begin{aligned} & \operatorname*{IP-lim}_{(\beta,\alpha) \in (\mathcal{F}^{(3)})^2_<} \int_Z \prod_{i=1}^t S_i(\alpha) f_i \prod_{i=1}^t S_i(\alpha \cup \beta) \overline{f_i} \, d\gamma \\ = & \operatorname*{IP-lim}_{(\beta,\alpha) \in (\mathcal{F}^{(3)})^2_<} \int_Z \prod_{i=1}^t S_i(\alpha) f_i \prod_{i=1}^t S_i(\alpha) S_i^{(2)}(\alpha, \beta) \left(S_i(\beta) \overline{f_i}\right) d\gamma = 0. \end{aligned} \tag{4.3}$$

If we write $S_i^{(\beta)}(\alpha) = S_i^{(2)}(\alpha, \beta)$ then the maps $\beta \rightarrow S_i^{(\beta)}$, $0 \leq i \leq t$, are integral polynomial forms on $\mathbf{PE}_{\leq d}(1)$. Furthermore, our earlier observations imply that for all $\beta \in \mathcal{F}^{(2)}$, one has $w(S_i^{(\beta)}) < w(S_i)$, $1 \leq i \leq t$.

By Lemma 4.9 there exists a subgroup $J < \mathbf{PE}_{\leq d}(1)$ with $G \cap J = \{I\}$ such that $[\mathbf{PE}_{\leq d}(1) : (G \oplus J)] < \infty$ and such that for any $U \in G$ and $W \in J$, $w(WU) \geq w(W)$. Since $[\mathbf{PE}_{\leq d}(1) : (G \oplus J)] < \infty$, by Remark 4.7 (ii) we may choose our refinement $\mathcal{F}^{(3)} \subset \mathcal{F}^{(2)}$ such that along $\mathcal{F}^{(3)}$ all the $S_i^{(\beta)}$ are integral polynomial forms on $G \oplus J$, so that in particular

$$S_i^{(\beta)} = W_i^{(\beta)} U_i^{(\beta)},$$

where $U_i^{(\beta)}$ are integral polynomial forms on G and $W_i^{(\beta)}$ are integral polynomial forms on J, $0 \leq i \leq t$. By the properties ascribed to J earlier,

$$w(S_i) > w(S_i^{(\beta)}) = w(W_i^{(\beta)} U_i^{(\beta)}) \geq w(W_i^{(\beta)}), \;\; 1 \leq i \leq t, \; \beta \in \mathcal{F}^{(3)}.$$

Since multiplication by a polynomial expression of lesser weight cannot affect the conjugacy class of S_i it follows that $S_i \sim S_i W_i^{(\beta)}$, $1 \leq i \leq t$, $\beta \in \mathcal{F}^{(3)}$.

Using Proposition 2.13, we may further require of $\mathcal{F}^{(3)}$ that

$$W_i^{(\beta)} \notin S_j S_i^{-1} G, \;\; 1 \leq i < j \leq t, \beta \in \mathcal{F}^{(3)} \tag{4.4}$$

and

$$W_i^{(\beta)} (W_j^{(\beta)})^{-1} \notin S_j S_i^{-1} G, \;\; 1 \leq i < j \leq t, \; \beta \in \mathcal{F}^{(3)}. \tag{4.5}$$

What enables us to do this is Theorem 2.15. For fixed i and j with $1 \leq i < j \leq t$, it is impossible to choose an IP ring along which $W_i^{(\beta)}$ (or $W_i^{(\beta)}(W_i^{(\beta)})^{-1}$) is *always* in $S_j S_i^{-1} G$. Hence Hindman's theorem allows us to choose a subring along which they are *never* in there. Similarly, we may require of $\mathcal{F}^{(3)}$ that for all i, $1 \leq i \leq t$, either $W_i^{(\beta)} = I$ for all $\beta \in \mathcal{F}^{(3)}$ or $W_i^{(\beta)} \neq I$ for all $\beta \in \mathcal{F}^{(3)}$. We now permute the indices $\{1, \cdots, t\}$ so that the indices i for which $W_i^{(\beta)} = I$ along $\mathcal{F}^{(3)}$ come first; that is there exists w with $0 \leq w \leq t$ such that $W_i^{(\beta)} = I$ if and only if $1 \leq i \leq w$. At the same time, we will require (for bookkeeping purposes and depending on whether the minimal weight occurs in the first or second group) that either S_1 or S_{w+1} has minimal weight among the S_i's. By (4.4) and (4.5), for any $\beta \in \mathcal{F}^{(3)}$ the set

$$A^{(\beta)} = \{S_1, \cdots, S_t, S_{w+1} W_{w+1}^{(\beta)}, \cdots, S_t W_t^{(\beta)}\}$$

is a mixing set, and furthermore this set has the same weight matrix as A (since $S_i \sim S_i W_i^{(\beta)}$, $w + 1 \leq i \leq t$). Assuming the limits exist, as Theorem 2.16 allows us to do (since there are countably many expression involved), we may rewrite (4.3) as

$$\begin{aligned} \underset{(\beta,\alpha)\in(\mathcal{F}^{(3)})^2_<}{\text{IP-lim}} \int_Z \prod_{i=1}^{w} S_i(\alpha)\Big(f_i \, U_i^{(\beta)}(\alpha)\big(S_i(\beta)\overline{f_i}\big)\Big) \\ \prod_{i=w+1}^{t} S_i(\alpha) f_i \; S_i(\alpha) W_i^{(\beta)}(\alpha)\Big(U_i^{(\beta)}(\alpha)\big(S_i(\beta)\overline{f_i}\big)\Big) \, d\gamma. \end{aligned} \tag{4.6}$$

Recall that either S_1 or S_{w+1} has minimal weight among the S_i's. Hence there are two cases to consider. We will confine ourselves to the case that S_1 has the minimal weight. (The other case is similar and even a bit easier, hence we shall omit it.) For any $\beta \in \mathcal{F}^{(3)}$, the set

$$S_1^{-1} A^{(\beta)} = \{S_2 S_1^{-1}, \cdots, S_t S_1^{-1}, S_{w+1} W_{w+1}^{(\beta)} S_1^{-1}, \cdots, S_t W_t^{(\beta)} S_1^{-1}\}$$

is a mixing set which precedes $A^{(\beta)}$ and hence precedes A. The reason for this is that S_1 is of minimal weight, and multiplying throughout by S_1^{-1} will decrease the weight of every expression which is equivalent to S_1, while failing to change the weight of the other expressions, or for that matter the conjugacy of any two other expressions. Hence the resulting family has one less conjugacy class at the S_1 weight and the same number at greater weights as the origial family. (This is the only point in the proof of Theorem 4.10 in which we create a family of lesser weight matrix from an existing family and is the core of the inductive technique.)

By hypothesis, (4) holds for the family $S_1^{-1}A^{(\beta)}$. Utilizing this fact, and noticing that in the limit below $\alpha \to \infty$ for each fixed β, (4.6) may be written as

$$\begin{aligned}
&\operatorname*{IP-lim}_{(\beta,\alpha)\in(\mathcal{F}^{(3)})^2_<} \int_Z \Big(f_1\, U_1^{(\beta)}(\alpha)\big(S_1(\beta)\overline{f_1}\big)\Big) \prod_{i=2}^{w} S_iS_1^{-1}(\alpha)\Big(f_i\, U_i^{(\beta)}(\alpha)\big(S_i(\beta)\overline{f_i}\big)\Big) \\
&\qquad \prod_{i=w+1}^{t} S_iS_1^{-1}(\alpha) f_i\; S_iW_i^{(\beta)}S_1^{-1}(\alpha)\Big(U_i^{(\beta)}(\alpha)\big(S_i(\beta)\overline{f_i}\big)\Big)\, d\gamma \\
&= \operatorname*{IP-lim}_{(\beta,\alpha)\in(\mathcal{F}^{(3)})^2_<} \int_Z E\Big(f_1\, U_1^{(\beta)}(\alpha)\big(S_1(\beta)\overline{f_1}\big)|Y\Big) \prod_{i=2}^{w} S_iS_1^{-1}(\alpha) E\Big(f_i\, U_i^{(\beta)}(\alpha)\big(S_i(\beta)\overline{f_i}\big)|Y\Big) \\
&\qquad \prod_{i=w+1}^{t} S_iS_1^{-1}(\alpha)E(f_i|Y)\; S_iW_i^{(\beta)}S_1^{-1}(\alpha)E\Big(U_i^{(\beta)}(\alpha)\big(S_i(\beta)\overline{f_i}\big)|Y\Big)\, d\gamma.
\end{aligned}$$

Recall that $E(f_{i_0}|Y) \equiv 0$. If $w+1 \leq i_0 \leq t$, the above limit is zero (consider that for fixed β the expression tends to zero as $\alpha \to \infty$) and we are done. Otherwise, since we are assuming that $||f_i|| \leq 1$, $0 \leq i \leq t$, the limit is still not greater than

$$\begin{aligned}
&\operatorname*{IP-lim}_{(\beta,\alpha)\in(\mathcal{F}^{(3)})^2_<} \Big|\Big|E\Big(f_{i_0}\, S_{i_0}(\beta)\big(U_{i_0}^{(\beta)}(\alpha)\overline{f_{i_0}}\big)|Y\Big)\Big|\Big| \\
&= \operatorname*{IP-lim}_{(\beta,\alpha)\in(\mathcal{F}^{(3)})^2_<} \Big|\Big|E\Big(\big(U_{i_0}^{(\beta)}\big)^{(-1)} f_{i_0}\, S_{i_0}(\beta)\overline{f_{i_0}}|Y\Big)\Big|\Big| = 0.
\end{aligned}$$

(We have used Proposition 4.2.) We are done in the case where $E(f_{i_0}|Y) = 0$ for some i_0. We reduce the general case to this special case by employing the identity

$$\prod_{i=0}^{t} a_i - \prod_{i=0}^{t} b_i = (a_0 - b_0)\prod_{i=1}^{t} b_i + a_0(a_1 - b_1)\prod_{i=2}^{t} b_i + \cdots + \Big(\prod_{i=0}^{t-1} a_i\Big)(a_t - b_t) \quad (4.7)$$

under the integral, with $a_i = S_i(\alpha)f_i$ and $b_i = S_i(\alpha)E(f_i|Y)$, $0 \leq i \leq t$.

(1) $\to$ (2) As in the previous case, we may assume without loss of generality that for some i_0, $E(f_{i_0}|Y) \equiv 0$. Recall that we are assuming the validity of (1) for every family in every primitive extension having the same weight matrix as A. In particular, we may assume that (1) holds for the family $\tilde{A} = \{\tilde{T} : T \in A\}$, which is mixing on $Z \times_Y Z$ relative to Y along $\mathcal{F}^{(1)}$ and has the same weight matrix as A.

Thus:

$$\begin{aligned}
&\underset{\alpha\in\mathcal{F}^{(2)}}{\text{IP-lim}}\ \Big|\Big|E\Big(\prod_{i=0}^{t} S_i(\alpha)f_i|Y\Big)\Big|\Big|^2\\
&=\underset{\alpha\in\mathcal{F}^{(2)}}{\text{IP-lim}}\ \int\Big|\int\prod_{i=0}^{t} S_i(\alpha)f_i d\gamma_y\Big|^2\, d\nu(y)\\
&=\underset{\alpha\in\mathcal{F}^{(2)}}{\text{IP-lim}}\ \int\prod_{i=0}^{t} \tilde{S}_i(\alpha)(f_i\otimes f_i)\, d\tilde{\gamma}\\
&=\underset{\alpha\in\mathcal{F}^{(2)}}{\text{IP-lim}}\ \int\prod_{i=0}^{t} \tilde{S}_i(\alpha)E(f_i\otimes \overline{f_i}|Y)\, d\nu\\
&=\underset{\alpha\in\mathcal{F}^{(2)}}{\text{IP-lim}}\ \int\prod_{i=0}^{t} \tilde{S}_i(\alpha)E(f_i|Y)E(\overline{f_i}|Y)\, d\nu=0.
\end{aligned}$$

(2)$\rightarrow$(3) We may assume without loss of generality that $||f_i^{(j)}||_\infty\leq 1$, $0\leq i\leq t$, $0\leq j\leq s$. By Theorem 3.11 there exists a refinement $\mathcal{F}^{(2)}\subset\mathcal{F}^{(1)}$ such that $AP(Z,Y,G,\mathcal{F}^{(2)})$ is dense in $\mathrm{L}^2(Z,\mathcal{C},\gamma)$. We may further require of $\mathcal{F}^{(2)}$ that

$$\underset{\alpha\in\mathcal{F}^{(2)}}{\text{IP-lim}}\ \Big|\Big|E\Big(\prod_{i=0}^{t} S_i(\alpha)\big(\prod_{j=0}^{s} U_i^{(j)}(\alpha)f_i^{(j)}\big)|Y\Big)-\prod_{i=0}^{t} S_i(\alpha)E\Big(\prod_{j=0}^{s} U_i^{(j)}(\alpha)f_i^{(j)}|Y\Big)\Big|\Big|$$

exist. We will show that this limit is zero by showing that for any $\delta>0$ we have

$$\Big|\Big|E\Big(\prod_{i=0}^{t} S_i(\alpha)\big(\prod_{j=0}^{s} U_i^{(j)}(\alpha)f_i^{(j)}\big)|Y\Big)-\prod_{i=0}^{t} S_i(\alpha)E\Big(\prod_{j=0}^{s} U_i^{(j)}(\alpha)f_i^{(j)}|Y\Big)\Big|\Big|\leq\delta$$

for all α far enough out. Let $\delta>0$ be arbitrary and let $\frac{\delta}{6}>\epsilon>0$. By Remark 3.2 (iv), there exist functions $g_1,\cdots,g_l\in L^\infty(Z,\mathcal{C},\gamma)$ (we may assume that $||g_i||_\infty\leq 1$, $1\leq i\leq l$) and $\alpha_0\in\mathcal{F}^{(2)}$ such that for each $\alpha\in\mathcal{F}^{(2)}$, $\alpha>\alpha_0$, there exists $E(\alpha)\in\mathcal{B}$ with $\nu\big(E(\alpha)\big)<\frac{\epsilon}{t+1}$ having the property that for $0\leq i\leq t$, $0\leq j\leq s$, and $y\in E(\alpha)^c$, there exists $v=v(i,j,\alpha,y)$ with $1\leq v\leq l$ such that

$$\Big|\Big|U_i^{(j)}(\alpha)f_i^{(j)}-g_{v(i,j,\alpha,y)}\Big|\Big|_y<\frac{\epsilon}{(s+1)(t+1)}.$$

Suppose that a_j, $b_j\in L^\infty(Z,\mathcal{C},\gamma_y)$, $0\leq j\leq s$. Then

$$\begin{aligned}
&\Big|\Big|\prod_{j=0}^{s} a_j-\prod_{j=0}^{s} b_j\Big|\Big|_y\\
=&\Big|\Big|(a_0-b_0)b_1\cdots b_l+a_0(a_1-b_1)b_2\cdots b_l+\cdots+a_0a_1\cdots a_{s-1}(a_s-b_s)\Big|\Big|_y\\
\leq&\sum_{j=0}^{s}||a_j-b_j||_y\Big(\sup_{0\leq i\leq s}\ \{||a_i||_\infty,\ ||b_i||_\infty\}\Big)^s.
\end{aligned}$$

Employment of this inequality with $a_j=U_i^{(j)}f_i^{(j)}$ and $b_j=g_{v(i,j,\alpha,y)}$, $0\leq j\leq s$ gives us

$$\Big|\Big|\prod_{j=0}^{s} U_i^{(j)}f_i^{(j)}-\prod_{j=0}^{s} g_{v(i,j,\alpha,y)}\Big|\Big|_y\leq\frac{\epsilon}{t+1},\quad 0\leq i\leq t,\ \alpha>\alpha_0,\ y\in E(\alpha)^c. \tag{4.8}$$

Let $h_1, \cdots, h_N$ be an enumeration of all products of the form $\prod_{j=0}^{s} g_{n_j}$, where $1 \leq n_j \leq l$, $0 \leq j \leq s$. Then of course $||h_i||_\infty \leq 1$, $1 \leq i \leq N$. Let $u : \{0, 1, \cdots, t\} \times \{\alpha \in \mathcal{F}^{(2)} : \alpha > \alpha_0\} \times E(\alpha)^c \to \{1, \cdots, N\}$ be defined by

$$h_{u(i,\alpha,y)} = \prod_{j=0}^{s} g_{(v(i,j,\alpha,y))}. \tag{4.9}$$

Then

$$\Big|\Big| \prod_{i=0}^{t} S_i(\alpha) \Big(\prod_{j=0}^{s} U_i^{(j)}(\alpha) f_i^{(j)} \Big) - \prod_{i=0}^{t} S_i(\alpha) h_{u(i,\alpha,y)} \Big|\Big|_y \leq \epsilon$$

whenever $\alpha > \alpha_0$ and

$$y \notin D(\alpha) = \bigcup_{i=0}^{t} S_i(\alpha)^{-1} E(\alpha).$$

Note that $\nu\big(D(\alpha)\big) < \epsilon$. Thus,

$$\Big|\Big| \prod_{i=0}^{t} S_i(\alpha) \Big(\prod_{j=0}^{s} U_i^{(j)}(\alpha) f_i^{(j)} \Big) - \prod_{i=0}^{t} S_i(\alpha) h_{u(i,\alpha,y)} \Big|\Big| < 2\epsilon < \frac{\delta}{3}. \tag{4.10}$$

On the other hand, by (4.8) and (4.9) we have

$$\Big| E\Big(\prod_{j=0}^{s} U_i^{(j)} f_i^{(j)} | Y \Big)(y) - E\Big(h_{u(i,\alpha,y)} | Y \Big)(y) \Big| \leq \frac{\epsilon}{t+1}$$

for $\alpha > \alpha_0$, $y \in E(\alpha)^c$ and $0 \leq i \leq t$. Again using identity (4.7) and recalling that $\nu(E(\alpha)) < \frac{\epsilon}{t+1}$, we have

$$\Big|\Big| \prod_{i=0}^{t} S_i(\alpha) E\Big(\prod_{j=0}^{s} U_i^{(j)}(\alpha) f_i^{(j)} | Y \Big) - \prod_{i=0}^{t} S_i(\alpha) E\big(h_{u(i,\alpha,y)} | Y \big) \Big|\Big| < 2\epsilon < \frac{\delta}{3}, \quad \alpha > \alpha_0. \tag{4.11}$$

However, since (2) is valid for the family $\{S_1, \cdots, S_t\}$, and $(h_i)_{i=1}^N$ is a finite set, we have

$$\begin{aligned}
&\operatorname*{IP-lim}_{\alpha \in \mathcal{F}^{(2)}} \Big|\Big| \prod_{i=0}^{t} S_i(\alpha) h_{u(i,\alpha,y)} - \prod_{i=0}^{t} S_i(\alpha) E(h_{u(i,\alpha,y)} | Y) \Big|\Big|^2 \\
&= \operatorname*{IP-lim}_{\alpha \in \mathcal{F}^{(2)}} \int \Big|\Big| \prod_{i=0}^{t} S_i(\alpha) h_{u(i,\alpha,y)} - \prod_{i=0}^{t} S_i(\alpha) E(h_{u(i,\alpha,y)} | Y) \Big|\Big|_y^2 \, d\nu \\
&\leq \operatorname*{IP-lim}_{\alpha \in \mathcal{F}^{(2)}} \int \sum_{j=1}^{N} \Big|\Big| \prod_{i=0}^{t} S_i(\alpha) h_j - \prod_{i=0}^{t} S_i(\alpha) E(h_j | Y) \Big|\Big|_y^2 \, d\nu \\
&= \operatorname*{IP-lim}_{\alpha \in \mathcal{F}^{(2)}} \sum_{j=1}^{N} \Big|\Big| \prod_{i=0}^{t} S_i(\alpha) h_j - \prod_{i=0}^{t} S_i(\alpha) E(h_j | Y) \Big|\Big|^2 = 0.
\end{aligned}$$

Hence, for α far enough out,

$$\Big|\Big| \prod_{i=0}^{t} S_i(\alpha) h_{u(i,\alpha,y)} - \prod_{i=0}^{t} S_i(\alpha) E(h_{u(i,\alpha,y)} | Y) \Big|\Big|^2 < \frac{\delta}{3}.$$

Combining this with inequalities (4.10) and (4.11) gives, as required,

$$\Big|\Big|E\Big(\prod_{i=0}^{t} S_i(\alpha)\big(\prod_{j=0}^{s} U_i^{(j)}(\alpha)f_i^{(j)}\big)|Y\Big) - \prod_{i=0}^{t} S_i(\alpha)E\Big(\prod_{j=0}^{s} U_i^{(j)}(\alpha)f_i^{(j)}|Y\Big)\Big|\Big| \leq \delta.$$

(3) → (4) Obvious. Strong convergence implies weak convergence.

□

Theorem 4.10, which deals with a mixing set in $\mathbf{PE}(1)$, is not quite what we need for the proof of Theorem 1.3, in which the polynomial expressions come from $\mathbf{PE}(m)$. This fact motivates the following definition and Theorem 4.12 to come.

Definition 4.11 Suppose that $(Z, \mathcal{C}, \gamma, \Omega)$ is a primitive extension of $(Y, \mathcal{B}, \nu, \Omega)$ along $\mathcal{F}^{(1)}$ with compact part G and suppose we are given a non-identity polynomial expression

$$S(\alpha_1, \cdots, \alpha_m) \in \mathbf{PE}_{\leq d}(m).$$

Write, in the standard way,

$$S(\alpha_1, \cdots, \alpha_m) = W(\alpha_1)W^{(\alpha_1)}(\alpha_2)W^{(\alpha_1,\alpha_2)}(\alpha_3)\cdots W^{(\alpha_1,\cdots,\alpha_{m-1})}(\alpha_m),$$

where $W \in \mathbf{PE}_{\leq d}(1)$ and $W^{(\alpha_1)}, \cdots, W^{(\alpha_1,\cdots,\alpha_{m-1})}$ are integral polynomial forms on $\mathbf{PE}_{\leq d}(1)$. We will say that $S(\alpha_1, \cdots, \alpha_m)$ is *mixing on Z relative to Y along $\mathcal{F}^{(1)}$* if there exists $\alpha_0 \in \mathcal{F}^{(1)}$ with the property that whenever $(\alpha_1, \cdots, \alpha_{m-1}) \in (\mathcal{F}^{(1)})^{m-1}_{<}$ with $\alpha_1 > \alpha_0$ it is the case that W, $W^{(\alpha_1)}, \cdots, W^{(\alpha_1,\cdots,\alpha_{m-1})}$ are all either the identity element of $\mathbf{PE}_{\leq d}(1)$ or are mixing on Z relative to Y along $\mathcal{F}^{(1)}$.

A family of polynomial expressions

$$\{S_1, \cdots, S_t\} \subset \mathbf{PE}_{\leq d}(m)$$

will be called a mixing set on Z relative to Y along $\mathcal{F}^{(1)}$ if S_i and $S_iS_j^{-1}$ are mixing on Z relative to Y along $\mathcal{F}^{(1)}$, $1 \leq i \neq j \leq t$.

Theorem 4.12 Suppose that $(Z, \mathcal{C}, \gamma, \Omega)$ is a primitive extension of $(Y, \mathcal{B}, \nu, \Omega)$ along $\mathcal{F}^{(1)}$ with compact part G and that $\{S_1, \cdots, S_t\} \subset \mathbf{PE}_{\leq d}(m)$ is mixing on Z relative to Y along $\mathcal{F}^{(1)}$. Let $S_0 = I$. If $f_i \in L^\infty(Z, \mathcal{C}, \gamma)$, $0 \leq i \leq t$, there exists a refinement $\mathcal{F}^{(2)} \subset \mathcal{F}^{(1)}$ with the property that

$$\operatorname*{IP-lim}_{(\alpha_1,\cdots,\alpha_m)\in(\mathcal{F}^{(2)})^m_<} \Big|\Big|E\Big(\prod_{i=0}^{t} S_i(\alpha_1, \cdots, \alpha_m)f_i|Y\Big) - \prod_{i=0}^{t} S_i(\alpha_1, \cdots, \alpha_m)E(f_i|Y)\Big|\Big| = 0.$$

Proof. We use induction on m. The case $m = 1$ is just Theorem 4.10 (2). Assume that the statement is true when m is replaced by $m - 1$. Without loss of generality we will again suppose that $||f_i||_\infty \leq 1$, $0 \leq i \leq t$ and that $E(f_{i_0}|Y) = 0$ for some i_0. We will show that for every $\delta > 0$ and every refinement $\mathcal{F}^{(2)} \subset \mathcal{F}^{(1)}$ there exists $(\alpha_1, \cdots, \alpha_m) \in (\mathcal{F}^{(2)})^m_<$ satisfying

$$\Big|\Big|E\Big(\prod_{i=0}^{t} S_i(\alpha_1, \cdots, \alpha_m)f_i|Y\Big)\Big|\Big| < \delta. \tag{4.12}$$

This will be sufficient by Theorem 2.14. Suppose then that $\mathcal{F}^{(2)} \subset \mathcal{F}^{(1)}$ is an arbitrary refinement. Write

$$S_i(\alpha_1, \cdots, \alpha_m) = W_i(\alpha_1) W_i^{(\alpha_1)}(\alpha_2) W_i^{(\alpha_1,\alpha_2)}(\alpha_3) \cdots W_i^{(\alpha_1,\cdots,\alpha_{m-1})}(\alpha_m).$$

Consider the set of polynomial forms

$$\mathcal{W} = \{W_i^{(\alpha_1,\cdots,\alpha_{m-1})} : 1 \leq i \leq t\}.$$

As the polynomial forms $W_i^{(\alpha_1,\cdots,\alpha_{m-1})}$ need not be distinct, $\mathcal{W}$ may have less than t elements. Reindexing if necessary we will assume that the distinct ones are represented in indices $1, \cdots, b$. There exists $\alpha_0 \in \mathcal{F}^{(2)}$ such that whenever $(\alpha_1, \cdots, \alpha_{m-1}) \in (\mathcal{F}^{(2)})^{m-1}_<$ the members of $\mathcal{W}$, evaluated at $(\alpha_1, \cdots, \alpha_{m-1})$, form a mixing set in $\mathbf{PE}_{\leq d}(1)$. Furthermore, for each j, $1 \leq j \leq b$, the set

$$\begin{aligned} A_j = \{S(\alpha_1, \cdots, \alpha_{m-1}) = W_i(\alpha_1) W_i^{(\alpha_1)}(\alpha_2) W_i^{(\alpha_1,\alpha_2)}(\alpha_3) \cdots \\ W_i^{(\alpha_1,\cdots,\alpha_{m-2})}(\alpha_{m-1}) :\ 1 \leq i \leq t,\ W_i^{(\alpha_1,\cdots,\alpha_{m-1})} = W_j^{(\alpha_1,\cdots,\alpha_{m-1})}\} \end{aligned}$$

is mixing on Z relative to Y along $\mathcal{F}^{(1)}$ in accordance with Definition 4.11. Notice that

$$\{S_1, \cdots, S_t\} = \bigcup_{j=1}^{b} \{S(\alpha_1, \cdots, \alpha_{m-1}) W_j^{(\alpha_1,\cdots,\alpha_{m-1})}(\alpha_m) : S \in A_j\}.$$

Reindex the f_i's in the manner suggested by the previous display (namely reindex f_i as $f_{S,j}$ where $S \in A_j$ and $S_i(\alpha_1, \cdots, \alpha_m) = S(\alpha_1, \cdots, \alpha_{m-1}) W_j^{(\alpha_1,\cdots,\alpha_{m-1})}(\alpha_m)$). We can rewrite (4.12) as

$$\Big|\Big|E\Big(\prod_{j=1}^{b} W_j^{(\alpha_1,\cdots,\alpha_{m-1})}(\alpha_m)\big(\prod_{S \in A_j} S(\alpha_1, \cdots, \alpha_{m-1}) f_{S,j}\big)|Y\Big)\Big|\Big| < \delta.$$

Recall that $E(f_{i_0}|Y) \equiv 0$ for some i_0. Let S_0 and j_0 be the new indices for f_{i_0}, so that $E(f_{S_0,j_0}|Y) = 0$. By the induction hypothesis and the fact that $A_{j_0} \subset \mathbf{PE}_{\leq d}(m-1)$ is a mixing set, there exists $(\alpha_1, \cdots, \alpha_{m-1}) \in (\mathcal{F}^{(2)})^{m-1}_<$ such that

$$\Big|\Big|E\Big(\prod_{S \in A_{j_0}} S(\alpha_1, \cdots, \alpha_{m-1}) f_{S,j_0}|Y\Big)\Big|\Big| < \frac{\delta}{2}.$$

Since we are assuming that all of our functions are bounded by 1, this gives

$$\Big|\Big|\prod_{j=1}^{b} W_j^{(\alpha_1,\cdots,\alpha_{m-1})}(\alpha_m) E\Big(\prod_{S \in A_j} S(\alpha_1, \cdots, \alpha_{m-1}) f_{S,j}|Y\Big)\Big|\Big| < \frac{\delta}{2} \tag{4.13}$$

for all $\alpha_m \in \mathcal{F}^{(2)}$ satisfying $\alpha_m > \alpha_{m-1}$.

On the other hand, for this fixed choice of $(\alpha_1, \cdots, \alpha_{m-1})$, by Theorem 4.10 (2) and the fact that $\{W_1^{(\alpha_1,\cdots,\alpha_{m-1})}, \cdots, W_b^{(\alpha_1,\cdots,\alpha_{m-1})}\}$ is a mixing set there exists some $\alpha_m \in \mathcal{F}^{(2)}$ with $\alpha_m > \alpha_{m-1}$ such that

$$\begin{aligned} \Big|\Big|E\Big(\prod_{j=1}^{b} W_j^{(\alpha_1,\cdots,\alpha_{m-1})}(\alpha_m)\big(\prod_{S \in A_j} S(\alpha_1, \cdots, \alpha_{m-1}) f_{S,j}\big)|Y\Big) \\ - \prod_{j=1}^{b} W_j^{(\alpha_1,\cdots,\alpha_{m-1})}(\alpha_m) E\Big(\prod_{S \in A_j} S(\alpha_1, \cdots, \alpha_{m-1}) f_{S,j}|Y\Big)\Big|\Big| < \frac{\delta}{2}. \end{aligned} \tag{4.14}$$

(4.13) and (4.14) combine to give (4.12), as desired.

□

We are nearly to the point where we can start directly in with the proof of Theorem 1.3. The two main tools will be Theorems 4.12 and 2.12. We do have one more piece of business (of an algebraical nature) to take care of, however, before it will be possible to use Theorem 2.12. The reader is asked now to recall Definitions 2.3 and 2.11. The content of the following proposition is that in the case of a primitive extension we may select an IP-ring along which the "partial derivatives" of polynomial expressions with compactness properties have compactness properties as well.

Proposition 4.13 Suppose $(Z,\mathcal{C},\gamma,\Omega)$ is a primitive extension of $(Y,\mathcal{B},\nu,\Omega)$ along $\mathcal{F}^{(1)}$ with compact part G. Then for some refinement $\mathcal{F}^{(2)}\subset\mathcal{F}^{(1)}$, the pair $(G,\mathcal{F}^{(2)})$ is balanced.

Proof. Throughout the course of this proof we will be using equation (2.8):

$$T(\alpha_1^{(1)}\cup\alpha_1^{(2)}\cup\cdots\cup\alpha_1^{(a_1)},\alpha_2^{(1)}\cup\alpha_2^{(2)}\cup\cdots\cup\alpha_2^{(a_2)},\cdots,\alpha_m^{(1)}\cup\alpha_m^{(2)}\cup\cdots\cup\alpha_m^{(a_m)})$$
$$=\prod_{\{\beta_i^{(1)},\cdots,\beta_i^{(b_i)}\}\subset\{\alpha_i^{(1)},\cdots,\alpha_i^{(a_i)}\},\ 1\le i\le m} T^{(b_1,\cdots,b_m)}(\beta_1^{(1)},\beta_1^{(2)},\cdots,\beta_1^{(b_1)},$$
$$\beta_2^{(1)},\beta_2^{(2)},\cdots,\beta_2^{(b_2)},\cdots,\beta_m^{(1)},\beta_m^{(2)},\cdots,\beta_m^{(b_m)}).\tag{2.8}$$

According to Lemma 4.9, there exists a subgroup $J<\mathbf{PE}_{\le d}(1)$ with $G\cap J=\{I\}$ and with $[\mathbf{PE}_{\le d}(1):G\oplus J]<\infty$. By Theorem 2.16 and Remark 4.7 (ii), we may select a refinement $\mathcal{F}^{(2)}\subset\mathcal{F}^{(1)}$ with the following property: for every integral polynomial form $S^{(\alpha_1,\cdots,\alpha_m)}$ on $\mathbf{PE}_{\le d}(1)$ there exists $\alpha_0\in\mathcal{F}^{(2)}$ such that either

(i) $S^{(\alpha_1,\cdots,\alpha_m)}\in G$ for all $(\alpha_1,\cdots,\alpha_m)\in(\mathcal{F}^{(2)})^m_<$ with $\alpha_0<\alpha_1$,

or

(ii) $S^{(\alpha_1,\cdots,\alpha_m)}\in(G\oplus J)\setminus G$ for all $(\alpha_1,\cdots,\alpha_m)\in(\mathcal{F}^{(2)})^m_<$ with $\alpha_0<\alpha_1$.

We must show that for every $m\in\mathbf{N}$, every $T(\alpha_1,\cdots,\alpha_m)\in\mathbf{PE}(G,\mathcal{F}^{(2)})$, and every $a_1,\cdots,a_m\in\mathbf{N}$ we have

$$T^{(a_1,\cdots,a_m)}\in\mathbf{PE}(G,\mathcal{F}^{(2)}).$$

The proof of this for general m is not hard, but it is somewhat complicated, at least notationally. For convenience, we will only show this for $m=1$ (to give a very general flavor) and for $m=3$ (all the ideas for the general case are present here).

First, consider the case $m=1$. That is, suppose that we have been given $T(\alpha)\in G$. We must show that $T^{(a)}\in\mathbf{PE}(G,\mathcal{F}^{(2)})$ for all $a\in\mathbf{N}$. We start by showing that $T^{(2)}(\alpha_1,\alpha_2)\in\mathbf{PE}(G,\mathcal{F}^{(2)})$. Writing $S^{(\alpha_1)}(\alpha_2)=T^{(2)}(\alpha_1,\alpha_2)$, $S^{(\alpha_1)}$ is an integral polynomial form on $\mathbf{PE}_{\le d}(1)$. Therefore, it satisfies either (i) or (ii). A moment's reflection makes it clear that what we are trying to show is in fact that it satisfies (i), so all we must do is show that it cannot satisfy (ii). Suppose then that it satisfies (ii). We will arrive at a contradiction.

Recall (see Definition 3.6) that we have a $\frac{Z}{Y}$-kernel H with the property that

$$\text{IP-}\lim_{\alpha\in\mathcal{F}^{(2)}}||U(\alpha)H-H||=0$$

for all $U \in G$ and

$$\underset{\alpha\in\mathcal{F}^{(2)}}{\text{IP-lim}}\ U(\alpha)H = 0 \tag{4.15}$$

weakly for all $U \in \big(\mathbf{PE}_{\leq d}(1) \setminus G\big)$. In particular,

$$\underset{\alpha\in\mathcal{F}^{(2)}}{\text{IP-lim}}\ \big|\big|U(\alpha)H - H\big|\big| = \sqrt{2}\big|\big|H\big|\big|,\ U \in \big(\mathbf{PE}_{\leq d}(1) \setminus G\big). \tag{4.16}$$

Consider the following inequality.

$$\begin{aligned}
&\big|\big|S^{(\alpha_1)}(\alpha_2)H - H\big|\big| \\
=&\big|\big|T^{(2)}(\alpha_1,\alpha_2)H - H\big|\big| \\
\leq&\big|\big|T^{(2)}(\alpha_1,\alpha_2)H - T(\alpha_1\cup\alpha_2)H\big|\big| + \big|\big|T(\alpha_1\cup\alpha_2)H - H\big|\big| \\
=&\big|\big|H - T(\alpha_1)T(\alpha_2)H\big|\big| + \big|\big|T(\alpha_1\cup\alpha_2)H - H\big|\big| \\
\leq&\big|\big|H - T(\alpha_2)H\big|\big| + \big|\big|T(\alpha_2)H - T(\alpha_1)T(\alpha_2)H\big|\big| + \big|\big|T(\alpha_1\cup\alpha_2)H - H\big|\big| \\
=&\big|\big|H - T(\alpha_2)H\big|\big| + \big|\big|H - T(\alpha_1)H\big|\big| + \big|\big|T(\alpha_1\cup\alpha_2)H - H\big|\big|.
\end{aligned} \tag{4.17}$$

The right hand side of (4.17) goes to zero as $(\alpha_1,\alpha_2) \in \big(\mathcal{F}^{(2)}\big)^2_<$ goes to infinity, hence the left hand side must go to zero as well. But if $S^{(\alpha_1)} \notin G$ then (4.15) implies that

$$\underset{\alpha\in\mathcal{F}^{(2)}}{\text{IP-lim}}\ \big|\big|S^{(\alpha_1)}(\alpha)H - H\big|\big| = \sqrt{2}\big|\big|H\big|\big|.$$

Hence (4.17) contradicts the assumption that $S^{(\alpha_1)} \notin G$ for all α_1 far enough out. Therefore we have established that $S^{(\alpha_1)} \in G$ for all α_1 far enough out, that is, $T^{(2)} \in \mathbf{PE}(G,\mathcal{F}^{(2)})$.

One goes on to show that $T^{(3)} \in \mathbf{PE}(G,\mathcal{F}^{(3)})$ by considering the inequality

$$\begin{aligned}
&\big|\big|T^{(3)}(\alpha_1,\alpha_2,\alpha_3)H - H\big|\big| \\
\leq&\big|\big|T^{(3)}(\alpha_1,\alpha_2,\alpha_3)H - T(\alpha_1\cup\alpha_2\cup\alpha_3)H\big|\big| + \big|\big|T(\alpha_1\cup\alpha_2\cup\alpha_3)H - H\big|\big| \\
=&\big|\big|H - T(\alpha_1)T(\alpha_2)T(\alpha_3)T^{(2)}(\alpha_1,\alpha_2)T^{(2)}(\alpha_1,\alpha_3)T^{(2)}(\alpha_2,\alpha_3)H\big|\big| \\
&\quad + \big|\big|T(\alpha_1\cup\alpha_2\cup\alpha_3)H - H\big|\big|.
\end{aligned}$$

The righthand side goes to zero by the previous case. The suggestion is that one gets $T^{(a)} \in \mathbf{PE}(G,\mathcal{F}^{(2)})$ for all $a > 3$ by induction. This is in fact so, though we leave out the details and move on to the case $m = 3$.

Suppose, then, that we are given

$$T(\alpha_1,\alpha_2,\alpha_3) \in \mathbf{PE}(G,\mathcal{F}^{(2)}).$$

We will show that $T^{(a_1,a_2,a_3)} \in \mathbf{PE}(G,\mathcal{F}^{(2)})$ for all non-negative integers a_1, a_2 and a_3. As was the case for $m = 1$, this is accomplished by induction, but before we can begin the inductive process we need to establish it separately for all choices of $a_i \in \{0,1\}$, $1 \leq i \leq 3$.

Since $T(\alpha_1,\alpha_2,\alpha_3) \in \mathbf{PE}(G,\mathcal{F}^{(2)})$, we may write (see Definition 2.3)

$$T(\alpha_1,\alpha_2,\alpha_3) = S(\alpha_1)S^{(\alpha_1)}(\alpha_2)S^{(\alpha_1,\alpha_2)}(\alpha_3), \tag{4.18}$$

where $S \in G$ and such that, for some $\alpha_0 \in \mathcal{F}^{(2)}$, we have $S^{(\alpha_1)} \in G$ and $S^{(\alpha_1,\alpha_2)} \in G$ whenever $(\alpha_1, \alpha_2) \in (\mathcal{F}^{(2)})^2_<$ with $\alpha_0 < \alpha_1$. On the other hand, we have

$$\begin{aligned}T(\alpha_1,\alpha_2,\alpha_3) =&T^{(1,0,0)}(\alpha_1)T^{(0,1,0)}(\alpha_2)T^{(1,1,0)}(\alpha_1,\alpha_2)T^{(0,0,1)}(\alpha_3)\\ &T^{(1,0,1)}(\alpha_1,\alpha_3)T^{(0,1,1)}(\alpha_2,\alpha_3)T^{(1,1,1)}(\alpha_1,\alpha_2,\alpha_3).\end{aligned} \tag{4.19}$$

(This is a consequence of equation (2.8).) Comparing (4.18) and (4.19), we see that

$$T^{(1,0,0)} = S \in G.$$

Again comparing (4.18) and (4.19), we have

$$S^{(\alpha_1)}(\alpha_2) = T^{(0,1,0)}(\alpha_2)T^{(1,1,0)}(\alpha_1,\alpha_2).$$

We may write $T^{(1,1,0)}(\alpha_1,\alpha_2) = U^{(\alpha_1)}(\alpha_2)$, where $U^{(\alpha_1)}$ is an integral polynomial form on $\mathbf{PE}_{\leq d}(1)$ which therefore satisfies either (i) or (ii). It follows that $U^{(\alpha_1)}$ is eventually contained in $G \oplus J$. (Recall that J is a subgroup of $\mathbf{PE}_{\leq d}(1)$ such that $J \cap G = \{I\}$ and $[\mathbf{PE}_{\leq d}(1) : G \oplus J] < \infty$.) We therefore have $U^{(\alpha_1)} = R^{(\alpha_1)}W^{(\alpha_1)}$, where $R^{(\alpha_1)}$ and $W^{(\alpha_1)}$ are integral polynomial forms with $R^{(\alpha_1)} \in G$ eventually and and $W^{(\alpha_1)} \in J$ eventually.

Recall that $S^{(\alpha_1)} \in G$ eventually. For $(\alpha_1,\alpha_2) \in \left(\mathcal{F}^{(2)}\right)^2_<$ far enough out we have

$$\left\{R^{(\alpha_1)}, R^{(\alpha_2)}, T^{(0,1,0)}R^{(\alpha_1)}W^{(\alpha_1)}(= S^{(\alpha_1)}), T^{(0,1,0)}R^{(\alpha_2)}W^{(\alpha_2)}(= S^{(\alpha_2)})\right\} \subset G.$$

Since G is a group, it follows that $W^{(\alpha_1)}(W^{(\alpha_2)})^{-1} \in G$. But $W^{(\alpha_1)}(W^{(\alpha_2)})^{-1} \in J$ as well, meaning that $W^{(\alpha_1)} = W^{(\alpha_2)}$ for all $(\alpha_1,\alpha_2) \in \left(\mathcal{F}^{(2)}\right)^2_<$ far enough out. Since any (eventually) constant integral polynomial form must be the identity, $W^{(\alpha_1)} = I$ and $T^{(1,1,0)}(\alpha_1,\cdot) = U^{(\alpha_1)} = R^{(\alpha_1)} \in G$ for $\alpha_1 \in \mathcal{F}^{(2)}$ far enough out. It follows that

$$T^{(1,1,0)} \in \mathbf{PE}(G, \mathcal{F}^{(2)}). \tag{4.21}$$

This also tells us that $T^{(0,1,0)} = S^{(\alpha_1)}\left(U^{(\alpha_1)}\right)^{-1} \in G$. In particular,

$$T^{(0,1,0)} \in \mathbf{PE}(G, \mathcal{F}^{(2)}). \tag{4.22}$$

Comparing (4.18) and (4.19) yet again, we have

$$S^{(\alpha_1,\alpha_2)}(\alpha_3) = T^{(0,0,1)}(\alpha_3)T^{(1,0,1)}(\alpha_1,\alpha_3)T^{(0,1,1)}(\alpha_2,\alpha_3)T^{(1,1,1)}(\alpha_1,\alpha_2,\alpha_3).$$

We now employ a strategy very similar to the one we just used. Namely, write

$$\begin{aligned}T^{(1,0,1)}(\alpha_1,\alpha_3) &= U_1^{(\alpha_1)}(\alpha_3) = R_1^{(\alpha_1)}(\alpha_3)W_1^{(\alpha_1)}(\alpha_3),\\ T^{(0,1,1)}(\alpha_2,\alpha_3) &= U_2^{(\alpha_2)}(\alpha_3) = R_2^{(\alpha_2)}(\alpha_3)W_2^{(\alpha_2)}(\alpha_3),\\ T^{(1,1,1)}(\alpha_1,\alpha_2,\alpha_3) &= U_3^{(\alpha_1,\alpha_2)}(\alpha_3) = R_3^{(\alpha_1,\alpha_2)}(\alpha_3)W_3^{(\alpha_1,\alpha_2)}(\alpha_3),\end{aligned}$$

where the R's are integral polynomial forms eventually contained in G and the W's are integral polynomial forms eventually contained in J.

Then, for $(\alpha_1,\cdots,\alpha_4)\in(\mathcal{F}^{(2)})^4_<$ far enough out we have

$$\big(W_1^{(\alpha_1)}W_2^{(\alpha_2)}W_3^{(\alpha_1,\alpha_2)}\big)\big(W_1^{(\alpha_3)}W_2^{(\alpha_4)}W_3^{(\alpha_3,\alpha_4)}\big)^{-1}\in G.$$

Reasoning as before we may conclude that $W^{(\alpha_1,\alpha_2)}=W_1^{(\alpha_1)}W_2^{(\alpha_2)}W_3^{(\alpha_1,\alpha_2)}=I$. But $W_1^{(\alpha_1)}$, $W_2^{(\alpha_2)}$, and $W_3^{(\alpha_1,\alpha_2)}$ correspond to the decomposition of the polynomial form $W^{(\alpha_1,\alpha_2)}$ into the parts of it that depend on α_1, on α_2, and on both, respectively. It is easily seen that these parts must all be the identity if their product is to be the identity. Hence we must have

$$W_1^{(\alpha_1)}=W_2^{(\alpha_2)}=W_3^{(\alpha_1,\alpha_2)}=I.$$

We now may routinely arrive at

$$\{T^{(1,0,1)},T^{(0,1,1)},T^{(1,1,1)},T^{(0,0,1)}\}\subset\mathbf{PE}(G,\mathcal{F}^{(2)}). \tag{4.23}$$

(4.20-23) establish that $T^{(a_1,a_2,a_3)}\in\mathbf{PE}(G,\mathcal{F}^{(2)})$ whenever $\{a_1,a_2,a_3\}\subset\{0,1\}$. The induction may now proceed. First one uses the fact that $T^{(1,0,0)}\in G$ to start an inductive process that proves that $T^{(a_1,0,0)}\in\mathbf{PE}(G,\mathcal{F}^{(2)})$ for all $a_1\in\mathbf{N}$. This is exactly like the case $m=1$ above. Proceed the same way in the other two coordinates as well, concluding that $T^{(0,a_2,0)},T^{(0,0,a_3)}\in\mathbf{PE}(G,\mathcal{F}^{(2)})$ for all $a_2,a_3\in\mathbf{N}$.

The next step is to use the fact that $T^{(1,1,0)}\in\mathbf{PE}(G,\mathcal{F}^{(2)})$ to get that $T^{(a_1,a_2,0)}\in\mathbf{PE}(G,\mathcal{F}^{(2)})$ for all $a_1,a_2\in\mathbf{N}$. The induction here is on the sum a_1+a_2, again using inequalities (recall the case $m=1$ above) involving the kernel H. Equation (2.8) is crucial here. For example, when $a_1=2$ and $a_2=1$ we have the inequality

$$\begin{aligned}
&||T^{(2,1,0)}(\alpha_1^{(1)},\alpha_1^{(2)},\alpha_2)H-H||\\
\le&||T^{(2,1,0)}(\alpha_1^{(1)},\alpha_1^{(2)},\alpha_2)H-T^{(1,1,0)}(\alpha_1^{(1)}\cup\alpha_1^{(2)},\alpha_2)H||\\
&\quad+||T^{(1,1,0)}(\alpha_1^{(1)}\cup\alpha_1^{(2)},\alpha_2)H-H||\\
=&||H-T^{(2,0,0)}(\alpha_1^{(1)},\alpha_1^{(2)})T^{(1,0,0)}(\alpha_1^{(1)})T^{(1,0,0)}(\alpha_1^{(2)})T^{(0,1,0)}(\alpha_2)\\
&\quad T^{(1,1,0)}(\alpha_1^{(1)},\alpha_2)T^{(1,1,0)}(\alpha_1^{(2)},\alpha_2)H||+||T^{(1,1,0)}(\alpha_1^{(1)}\cup\alpha_1^{(2)},\alpha_2)H-H||.
\end{aligned} \tag{4.24}$$

We are using here the identity

$$\begin{aligned}
&T^{(1,1,0)}(\alpha_1^{(1)}\cup\alpha_1^{(2)},\alpha_2)\\
=&T^{(2,1,0)}(\alpha_1^{(1)},\alpha_1^{(2)},\alpha_2)T^{(2,0,0)}(\alpha_1^{(1)},\alpha_1^{(2)})T^{(1,0,0)}(\alpha_1^{(1)})\\
&\quad T^{(1,0,0)}(\alpha_1^{(2)})T^{(0,1,0)}(\alpha_2)T^{(1,1,0)}(\alpha_1^{(1)},\alpha_2)T^{(1,1,0)}(\alpha_1^{(2)},\alpha_2).
\end{aligned}$$

which is a case of equation (2.8).

(4.24) allows us to conclude from previous cases that

$$\operatorname*{IP-lim}_{(\alpha_1^{(1)},\alpha_1^{(2)},\alpha_2)\in(\mathcal{F}^{(2)})^3_<}||T^{(2,1,0)}(\alpha_1^{(1)},\alpha_1^{(2)},\alpha_2)H-H||=0.$$

Suppose now that $T^{(2,1,0)}$ satisfies (ii). Then for $(\alpha_1^{(1)},\alpha_1^{(2)})\in\big(\mathcal{F}^{(2)}\big)^2_<$ far enough out, $T^{(2,1,0)}(\alpha_1^{(1)},\alpha_1^{(2)},\cdot)\notin G$, which mean that

$$\operatorname*{IP-lim}_{\alpha_2\in\mathcal{F}^{(2)}}||T^{(2,1,0)}(\alpha_1^{(1)},\alpha_1^{(2)},\alpha_2)H-H||=\sqrt{2}||H||,$$

a contradiction. Therefore $T^{(2,1,0)} \in \mathbf{PE}(G, \mathcal{F}^{(2)})$.

One proceeds in the same manner with the other pairs of indices, then goes finally to triples $T^{(a_1,a_2,a_3)}$, $a_1, a_2, a_3 \in \mathbf{N}$ (by induction on $a_1+a_2+a_3$). The same sort of inequalities play a key role the whole way. This completes the proof (or at least a sketch of a proof) in the case $m = 3$. As we said, the case of general m is completely analogous, so we have established that $(G, \mathcal{F}^{(2)})$ is a balanced pair.

□

CHAPTER 5

COMPLETION OF THE PROOF

We are now ready for the final steps in the proof of Theorem 1.3, which we now restate.

Theorem 1.3 Let $k, r, d \in \mathbf{N}$ and fix IP-sets $(n_\alpha^{(i)})_{\alpha \in \mathcal{F}}$, $1 \leq i \leq k$. Let $\mathbf{PE}_{\leq d}(m)$ be as in Definition 1.1 and suppose that $(X, \mathcal{A}, \mu, \Omega)$ is a measure preserving system, where Ω is generated by commuting transformations $T_1, \cdots, T_r$. For every $A \in \mathcal{A}$ with $\mu(A) > 0$ and every $m, t \in \mathbf{N}$ there exist an IP-ring $\mathcal{F}^{(1)}$ and a number $a = a(A, m, t, d) > 0$ having the property that for every set of polynomial expressions $\{S_0, \cdots, S_t\} \subset \mathbf{PE}_{\leq d}(m)$ we have

$$\underset{(\alpha_1, \cdots, \alpha_m) \in (\mathcal{F}^{(1)})^m_<}{\text{IP-lim}} \mu\Big(\bigcap_{i=0}^{t} S_i(\alpha_1, \cdots, \alpha_m)^{-1} A\Big) \geq a.$$

Recall that $k, r \in \mathbf{N}$ and the IP-sets $(n_\alpha^{(i)})_{\alpha \in \mathcal{F}}$, though arbitrary, have been fixed since Section 2. Our plan is to verify that the conclusion of Theorem 1.3 holds for these fixed quantities and an arbitrary system $(X, \mathcal{A}, \mu, \Omega)$ for which the underlying space is separable. Having done this, it will be evident that the conclusion also holds for an arbitrary, possibly non-separable system. This is because the IP-ring $\mathcal{F}^{(1)}$ in the formulation is allowed to depend on the set A. Therefore, having chosen A, one may before choosing $\mathcal{F}^{(1)}$ replace $(X, \mathcal{A}, \mu, \Omega)$ by the separable factor generated by the iterates of A under the transformations of Ω.

Fix a *Lebesgue* (in particular, separable) system $(X, \mathcal{A}, \mu, \Omega)$, where Ω is generated by commuting measure preserving transformations $T_1, T_2, \cdots, T_r$. In Remark 5.4 we will show that, having established the result for Lebesgue systems, it follows for general separable systems. As we have mentioned, the proof of the validity of the conclusion to Theorem 1.3 for the system $(X, \mathcal{A}, \mu, \Omega)$ is accomplished by the method of exhaustion of the σ-algebra $\mathcal{A}$. In particular, we achieve it by passing the following property through a transfinite sequence of sub-σ-algebras of $\mathcal{A}$:

Definition 5.1 Suppose that $d \in \mathbf{N}$, $\mathcal{B} \subset \mathcal{A}$ is an Ω-invariant σ-algebra, and $\mathcal{F}^{(1)}$ is an IP-ring. The pair $(\mathcal{B}, \mathcal{F}^{(1)})$ is said to have the *IPPSZ property of degree* d if for every $A \in \mathcal{B}$ with $\mu(A) > 0$ and every $t, m \in \mathbf{N}$ there exists a number $a = a(A, m, t, d) > 0$ having the property that for every set of polynomial expressions $\{S_0, \cdots, S_t\} \subset \mathbf{PE}_{\leq d}(m)$ we have

$$\underset{(\alpha_1, \cdots, \alpha_m) \in (\mathcal{F}^{(1)})^m_<}{\text{IP-lim}} \mu\Big(\bigcap_{i=0}^{t} S_i(\alpha_1, \cdots, \alpha_m)^{-1} A\Big) \geq a.$$

The reader should convince himself that in order to show that the conclusion of Theorem 1.3 holds for the system $(X, \mathcal{A}, \mu, \Omega)$, it is sufficient to show that for some IP-ring $\mathcal{G}$, the pair $(\mathcal{A}, \mathcal{G})$ has the IPPSZ property of degree d. This is what we will show. Note however, that Definition 5.1 is tailored for a separable system $(X, \mathcal{A}, \mu, \Omega)$. We are not claiming that a similar property may be passed through a non-separable system. (The point here is that in Definition 5.1 the IP-ring does not depend on A, as it does in Theorem 1.3. Therefore in the non-separable case there is no way to ensure that all of the IP-limits under consideration even exist.)

The groundwork for the proof that $(\mathcal{A}, \mathcal{G})$ has the IPPSZ property for some IP-ring $\mathcal{G}$ has been laid. Let us now fix an IP-ring $\mathcal{F}^{(1)}$ and a number $d \in \mathbf{N}$. In order to use our exhaustion technique, we need the following partial order on the set of pairs $(\mathcal{B}, \mathcal{F}^{(2)})$, where $\mathcal{B} \subset \mathcal{A}$ is an Ω-invariant, complete (with respect to μ) sub-σ-algebra and $\mathcal{F}^{(2)} \subset \mathcal{F}^{(1)}$ is an IP-ring. If $\mathcal{B}_1$, $\mathcal{B}_2$ are complete, Ω-invariant sub-σ-algebras of $\mathcal{A}$, and $\mathcal{F}^{(2)}$, $\mathcal{F}^{(3)}$ are refinements of $\mathcal{F}^{(1)}$, we will write $(\mathcal{B}_1, \mathcal{F}^{(2)}) < (\mathcal{B}_2, \mathcal{F}^{(3)})$ if $\mathcal{B}_1 \subset \mathcal{B}_2$ properly (in the sense that $\mathcal{B}_1 \neq \mathcal{B}_2 \bmod 0$) and if there exists $\alpha_0 \in \mathcal{F}^{(3)}$ such that for each $\alpha \in \mathcal{F}^{(3)}$ with $\alpha > \alpha_0$ we have $\alpha \in \mathcal{F}^{(2)}$. (That is, $\mathcal{F}^{(3)} \subset \mathcal{F}^{(2)}$ asymptotically.)

Proposition 5.2 Among all pairs $(\mathcal{B}, \mathcal{F}^{(2)})$, where $\mathcal{F}^{(2)} \subset \mathcal{F}^{(1)}$ is a refinement and $\mathcal{B} \subset \mathcal{A}$ is a closed Ω-invariant σ-algebra, having the IPPSZ property of degree d, there exists a pair which is maximal with respect to the order described above.

Proof. We will use Zorn's Lemma. Suppose that $(\mathcal{B}_i, \mathcal{G}^{(i)})$ is a totally ordered chain of pairs having the IPPSZ property of degree d. We may, by separability of $(X, \mathcal{A}, \mu)$, assume that this chain is countable. (Indeed, it must be.) Let $\mathcal{C}$ be the μ-completion of the Ω-invariant σ-algebra generated by $\bigcup_i \mathcal{B}_i$. Then $\bigcup_i \mathcal{B}_i$ is an Ω-invariant algebra which is dense in $\mathcal{C}$. Let $\mathcal{F}^{(2)}$ be an IP-ring of the form $\mathcal{F}^{(2)} = FU(\{\beta_1, \beta_2, \cdots\})$, where $\beta_i \in \bigcap_{j=1}^{i} \mathcal{G}^{(j)}$ and $\beta_1 < \beta_2 < \cdots$. Notice that $\mathcal{F}^{(2)}$ is asymptotically contained in $\mathcal{F}_i$ for all $i \in \mathbf{N}$. Since $(X, \mathcal{A}, \mu)$ is separable, by passing to a subring if necessary we may assume that

$$\operatorname*{IP-lim}_{(\alpha_1, \cdots, \alpha_m) \in (\mathcal{F}^{(1)})^m_<} \mu\Big(\bigcap_{i=0}^{t} S_i(\alpha_1, \cdots, \alpha_m)^{-1} A\Big)$$

exists for every $A \in \mathcal{A}$ and every set $\{S_0, \cdots, S_t\} \subset \mathbf{PE}_{\leq d}(m)$. We claim that $(\mathcal{C}, \mathcal{F}^{(2)})$ has the IPPSZ property of degree d.

Let $(Z, \mathcal{C}, \gamma, \Omega)$ be the factor of $(X, \mathcal{A}, \mu, \Omega)$ determined by $\mathcal{C}$. Suppose now that $A \in \mathcal{C}$ with $\gamma(A) > 0$ and $t, m \in \mathbf{N}$. We can find a fixed number $i \in \mathbf{N}$ and a set $A_i \in \mathcal{B}_i$ such that

$$\gamma(A \triangle A_i) < \frac{\gamma(A)}{8(t+1)}.$$

Let $\mathcal{B} = \mathcal{B}_i$ and let $(Y, \mathcal{B}, \nu, \Omega)$ be the factor of $(Z, \mathcal{C}, \gamma, \Omega)$ determined by $\mathcal{B}$. Let $\gamma = \int \gamma_y \, d\nu(y)$ be the regular decomposition of γ over Y. We have

$$\gamma(A_i \setminus A) = \int_{A_i} 1 - \gamma_y(A) \, d\nu(y) < \frac{\gamma(A)}{8(t+1)}.$$

It follows that

$$\nu\Big(\{y \in A_i : 1 - \gamma_y(A) \geq \frac{1}{2(t+1)}\}\Big) < \frac{\gamma(A)}{4}.$$

Hence there exists a set $A_i' \in \mathcal{B}$ with $A_i' \subset A_i$ and $\nu(A_i') > 0$ such that for all $y \in A_i'$, $\gamma_y(A) > 1 - \frac{1}{2(t+1)}$.

Recall that $(\mathcal{B}, \mathcal{G}^{(i)})$ has the IPPSZ property of degree d. Let $a = a(A_i', m, t, d)$ be as guaranteed by Definition 5.1. We claim that for every set $\{S_0, \cdots, S_t\} \subset \mathbf{PE}_{\leq d}(1)$,

$$\operatorname*{IP-lim}_{(\alpha_1,\cdots,\alpha_m)\in(\mathcal{F}^{(2)})^m_<} \mu\Big(\bigcap_{i=0}^{t} S_i(\alpha_1,\cdots,\alpha_m)^{-1}A\Big) \geq \frac{1}{4}a.$$

This will be sufficient for the proof, as we can then put $a(A,m,t,d) = \frac{1}{4}a$. Suppose then to the contrary that for some set $\{S_0, \cdots, S_t\} \in \mathbf{PE}_{\leq d}(m)$ we have

$$\operatorname*{IP-lim}_{(\alpha_1,\cdots,\alpha_m)\in(\mathcal{F}^{(2)})^m_<} \mu\Big(\bigcap_{j=0}^{t} S_j(\alpha_1,\cdots,\alpha_m)^{-1}A\Big) < \frac{1}{4}a.$$

Since $\mathcal{F}^{(2)}$ is asymptotically contained in $\mathcal{G}^{(i)}$, we may pass to a subring $\mathcal{F}^{(3)} \subset \mathcal{G}^{(i)} \cap \mathcal{F}^{(2)}$ such that

$$\mu\Big(\bigcap_{j=0}^{t} S_j(\alpha_1,\cdots,\alpha_m)^{-1}A\Big) < \frac{1}{4}a$$

for all $(\alpha_1, \cdots, \alpha_m) \in (\mathcal{F}^{(3)})^m_<$.

On the other hand, by hypothesis we can find some $(\alpha_1, \cdots, \alpha_m) \in (\mathcal{F}^{(3)})^m_<$ (which we now fix) satisfying

$$\nu\Big(\bigcap_{j=0}^{t} S_j(\alpha_1,\cdots,\alpha_m)^{-1}A_i'\Big) > \frac{1}{2}a.$$

For $y \in \bigcap_{j=0}^t S_j(\alpha_1, \cdots, \alpha_m)^{-1}A_i'$, we have $S_j(\alpha_1, \cdots, \alpha_m)y \in A_j'$, $0 \leq j \leq t$, so that

$$\gamma_{S_j(\alpha_1,\cdots,\alpha_m)y}(A_i') > 1 - \frac{1}{2(t+1)}, \quad 0 \leq j \leq t.$$

This is the same as

$$\gamma_y\Big(S_j(\alpha_1,\cdots,\alpha_m)^{-1}A_i'\Big) > 1 - \frac{1}{2(t+1)}, \quad 0 \leq j \leq t.$$

It follows that

$$\gamma_y\Big(\bigcap_{j=0}^{t} S_j(\alpha_1,\cdots,\alpha_m)^{-1}A_i'\Big) > \frac{1}{2}, \; y \in \bigcap_{j=0}^{t} S_j(\alpha_1,\cdots,\alpha_m)^{-1}A_i'.$$

But $\nu\big(\bigcap_{j=0}^{l} S_j(\alpha_1, \cdots, \alpha_m)^{-1}A_i'\big) > \frac{1}{2}a$, whence

$$\gamma\Big(\bigcap_{j=0}^{t} S_j(\alpha_1,\cdots,\alpha_m)^{-1}A\Big) > \frac{1}{4}a,$$

a contradiction.

□

All that remains in order for the proof of Theorem 1.3 to be completed is to show that the IPPSZ property passes through primitive extensions, which we do now.

Theorem 5.3 Suppose that $\mathcal{B} \subset \mathcal{A}$ is a complete Ω-invariant proper sub-σ-algebra and that $\mathcal{F}^{(1)}$ is an IP-ring such that $(\mathcal{B}, \mathcal{F}^{(1)})$ has the IPPSZ property of degree d. Then there exists an Ω-invariant σ-algebra $\mathcal{C} \subset \mathcal{A}$, with $\mathcal{B} \subset \mathcal{C}$ properly ($\mathcal{B} \neq \mathcal{C}$ mod 0), and a refinement $\mathcal{F}^{(2)} \subset \mathcal{F}^{(1)}$ such that $(\mathcal{C}, \mathcal{F}^{(2)})$ has the IPPSZ property of degree d.

Proof. Let $(Y, \mathcal{B}, \nu, \Omega)$ be the factor of $(X, \mathcal{A}, \mu, \Omega)$ determined by $\mathcal{B}$. Theorem 3.15 guarantees that we can find a refinement $\mathcal{F}^{(2)} \subset \mathcal{F}^{(1)}$, a proper extension $(Z, \mathcal{C}, \gamma, \Omega)$ of $(Y, \mathcal{B}, \nu, \Omega)$ (which is a factor of $(X, \mathcal{A}, \mu, \Omega)$), and a subgroup $G < \mathbf{PE}_{\leq d}(1)$ such that $(Z, \mathcal{C}, \gamma, \Omega)$ is a primitive extension of $(Y, \mathcal{B}, \nu, \Omega)$ along $\mathcal{F}^{(2)}$ with compact part G. By Lemma 3.11 we may further require of $\mathcal{F}^{(2)}$ that $AP(Z, Y, G, \mathcal{F}^{(2)})$ be dense in $\mathrm{L}^2(Z, \mathcal{C}, \gamma)$. By Theorem 2.14 and the separability of $\mathcal{C}$ we may assume that for all $A \in \mathcal{C}$ and all finite sets $\{S_0, \cdots, S_t\} \subset \mathbf{PE}_{\leq d}(m)$ we have existence of the limit

$$\operatorname*{IP-lim}_{(\alpha_1, \cdots, \alpha_m) \in (\mathcal{F}^{(2)})^m_<} \mu\Big(\bigcap_{i=0}^{t} S_i(\alpha_1, \cdots, \alpha_m)^{-1} A\Big).$$

We claim that $(\mathcal{C}, \mathcal{F}^{(2)})$ has the IPPSZ property of degree d.

Suppose that $t, m \in \mathbf{N}$ and $A \in \mathcal{C}$ with $\mu(A) > 0$. Fix a number $a = a(A, m, t, d) > 0$ which will be determined presently. Let $\{S_0, \cdots, S_t\} \subset \mathbf{PE}_{\leq d}(m)$. (It is not practical for us to write down now what a is. However we will check when we do write it down that it does not depend on the set $\{S_0, \cdots, S_t\}$.)

What we have to show is that

$$\operatorname*{IP-lim}_{(\alpha_1, \cdots, \alpha_m) \in (\mathcal{F}^{(2)})^m_<} \mu\Big(\bigcap_{i=0}^{t} S_i(\alpha_1, \cdots, \alpha_m)^{-1} A\Big) \geq a.$$

Suppose that this is not the case. Then by Theorem 2.15 there exists an IP-ring $\mathcal{F}^{(3)} \subset \mathcal{F}^{(2)}$ such that

$$\mu\Big(\bigcap_{i=0}^{t} S_i(\alpha_1, \cdots, \alpha_m)^{-1} A\Big) < a$$

for every $(\alpha_1, \cdots, \alpha_m) \in \left(\mathcal{F}^{(3)}\right)^m_<$. We will contradict this by exhibiting a single m-tuple $(\alpha_1, \cdots, \alpha_m) \in (\mathcal{F}^{(3)})^m_<$ with

$$\gamma\Big(\bigcap_{i=0}^{t} S_i(\alpha_1, \cdots, \alpha_m)^{-1} A\Big) \geq a,$$

whereupon the proof will have been completed. We may assume without loss of generality (by multiplying throughout by S_0^{-1}) that $S_0 = I$.

As usual, we let $J < \mathbf{PE}_{\leq d}(1)$ be a subgroup with $J \cap G = \{I\}$ and with $[\mathbf{PE}_{\leq d}(1) : G \oplus J] < \infty$. For $i = 0, 1, \cdots, t$ write

$$S_i(\alpha_1, \cdots, \alpha_m) = U_i(\alpha_1) U_i^{(\alpha_1)}(\alpha_2) \cdots U_i^{(\alpha_1, \cdots, \alpha_{m-1})}(\alpha_m), \tag{5.1}$$

as in (2.1). Now let $S_i^{(j)} \in \mathbf{PE}_{\leq d}(1)$ be the portion of S_i that depends only on α_j, $1 \leq j \leq m$. In other words, $S_i^{(1)} = S_i^{(1,0,\cdots,0)}$, $S_i^{(2)} = S_i^{(0,1,0,\cdots,0)}, \cdots, S_i^{(m)} = S_i^{(0,\cdots,0,1)}$. Finally, we set

$$\begin{aligned} S_i^{(\alpha_1)} &= U_i^{(\alpha_1)}\big(S_i^{(2)}\big)^{-1}, \\ S_i^{(\alpha_1,\alpha_2)} &= U_i^{(\alpha_1,\alpha_2)}\big(S_i^{(3)}\big)^{-1}, \\ &\vdots \\ S_i^{(\alpha_1,\cdots,\alpha_{m-1})} &= U_i^{(\alpha_1,\cdots,\alpha_{m-1})}\big(S_i^{(m)}\big)^{-1}. \end{aligned}$$

This gives us the following alternate decomposition of S_i:

$$S_i(\alpha_1,\cdots,\alpha_m) = S_i^{(1)}(\alpha_1)\cdots S_i^{(m)}(\alpha_m)S_i^{(\alpha_1)}(\alpha_2)\cdots S_i^{(\alpha_1,\cdots,\alpha_{m-1})}(\alpha_m). \qquad (5.2)$$

The reason we prefer this decomposition to that given in (5.1) is that the polynomial forms appearing in (5.2), namely $S_i^{(\alpha_1)}, \cdots, S_i^{(\alpha_1,\cdots,\alpha_{m-1})}$, are *integral* polynomial forms on $\mathbf{PE}_{\leq d}(1)$. Of course $S_i^{(1)}, \cdots, S_i^{(m)}$ are fixed elements of $\mathbf{PE}_{\leq d}(1)$.

Let $\mathcal{W}$ be a set of coset representatives, containing I, for $\mathbf{PE}_{\leq d}(1)$ modulo G. We may write

$$S_i^{(j)} = R_i^{(j)}W_i^{(j)}, \quad 0 \leq i \leq t, \quad 1 \leq j \leq m,$$

where $R_i^{(j)} \in G$ and $W_i^{(j)} \in \mathcal{W}$. Since $[\mathbf{PE}_{\leq d}(1) : G \oplus J] < \infty$, according to Remark 4.7 (ii) we may pass to a suitable sub-IP-ring (which we continue to call $\mathcal{F}^{(3)}$) along which $S_i^{(\alpha_1,\cdots,\alpha_j)}$ is an integral polynomial form on $G \oplus J$, $1 \leq i \leq t$, $1 \leq j < m$. By Remark 4.7 (i) we may therefore write

$$S_i^{(\alpha_1,\cdots,\alpha_j)} = R_i^{(\alpha_1,\cdots,\alpha_j)}W_i^{(\alpha_1,\cdots,\alpha_j)}, \quad 0 \leq i \leq t, \quad 1 \leq j < m,$$

where $R_i^{(\alpha_1,\cdots,\alpha_j)}$ is an integral polynomial form on G and $W_i^{(\alpha_1,\cdots,\alpha_j)}$ is an integral polynomial form on J.

Let

$$R_i(\alpha_1,\cdots,\alpha_m) = R_i^{(1)}(\alpha_1)\cdots R_i^{(m)}(\alpha_m)R_i^{(\alpha_1)}(\alpha_2)\cdots R_i^{(\alpha_1,\cdots,\alpha_{m-1})}(\alpha_m), \quad 0 \leq i \leq t.$$

Obviously $R_i \in \mathbf{PE}(G, \mathcal{F}^{(3)})$, $0 \leq i \leq t$. Now let

$$W_i(\alpha_1,\cdots,\alpha_m) = W_i^{(1)}(\alpha_1)\cdots W_i^{(m)}(\alpha_m)W_i^{(\alpha_1)}(\alpha_2)\cdots W_i^{(\alpha_1,\cdots,\alpha_{m-1})}(\alpha_m).$$

Then

$$S_i(\alpha_1,\cdots,\alpha_m) = R_i(\alpha_1,\cdots,\alpha_m)W_i(\alpha_1,\cdots,\alpha_m), \quad 0 \leq i \leq t.$$

We claim that there exists some refinement $\mathcal{G}$ of $\mathcal{F}^{(3)}$ having the property that if $W_i(\alpha_1,\cdots,\alpha_m)$ is not the identity then it is mixing on $(Z,\mathcal{C},\gamma,\Omega)$ relative to $(Y,\mathcal{B},\nu,\Omega)$ along $\mathcal{G}$, $1 \leq i \leq t$. Indeed, suppose this were not the case. Then by an application of the Milliken-Taylor theorem to the definition of mixing we would be able to choose some i, j, and a refinement $\mathcal{G} \subset \mathcal{F}^{(3)}$ such that $W_i^{(j)}W_i^{(\alpha_1,\cdots,\alpha_{j-1})}$ is a non-identity polynomial form on G (notice that this polynomial expression when evaluated at α_j is the part of $W_i(\alpha_1,\cdots,\alpha_m)$ which depends only on $\alpha_1,\cdots,\alpha_j$ but not on $\alpha_1,\cdots,\alpha_{j-1}$ alone). This implies that $W_i^{(\alpha_1,\cdots,\alpha_{j-1})}$ is a non-identity polynomial form (remember $W_i^{(j)} \in \mathcal{W}$, so if it is in G then it is the identity). If $(\alpha_1, \cdots, \alpha_{j-1}), (\beta_1, \cdots, \beta_{j-1}) \in (\mathcal{G})_<^{j-1}$ then $W_i^{(\alpha_1,\cdots,\alpha_{j-1})}(W_i^{(\beta_1,\cdots,\beta_{j-1})})^{-1} \in G$. On the other hand since $W_i^{(\alpha_1,\cdots,\alpha_{j-1})}$ is a polynomial expression on J, and J is a group, $W_i^{(\alpha_1,\cdots,\alpha_{j-1})}(W_i^{(\beta_1,\cdots,\beta_{j-1})})^{-1} \in J$. But $G \cap J = \{I\}$, so $W_i^{(\alpha_1,\cdots,\alpha_{j-1})} = W_i^{(\beta_1,\cdots,\beta_{j-1})}$, which since $(\alpha_1, \cdots, \alpha_{j-1})$ and $(\beta_1, \cdots, \beta_{j-1})$ are arbitrary elements of $\mathcal{G}_<^{j-1}$, implies that $W_i^{(\alpha_1,\cdots,\alpha_{j-1})}$ is a constant non-identity integral polynomial form (an impossibility). This contradiction establishes the claim.

Hence we may, by passing if necessary to a subring, assume that $\mathcal{F}^{(3)}$ has the property that if $W_i(\alpha_1, \cdots, \alpha_m)$ is not the identity then it is mixing on $(Z, \mathcal{C}, \gamma, \Omega)$ relative to $(Y, \mathcal{B}, \nu, \Omega)$ along $\mathcal{F}^{(3)}$, $1 \le i \le t$. A similar argument can be used to show that we may also assume of $\mathcal{F}^{(3)}$ that if $W_i(\alpha_1, \cdots, \alpha_m)W_j(\alpha_1, \cdots, \alpha_m)^{-1}$ is not the identity then it is mixing on $(Z, \mathcal{C}, \gamma, \Omega)$ relative to Y along $\mathcal{F}^{(3)}$, $1 \le i, j \le t$. We may by renumbering (if necessary, and leaving $S_0 = I$) assume that the distinct, non-identity members of the set $\{W_0, W_1, \cdots, W_t\}$ are represented in the indices $\{1, \cdots, v\}$ (with $W_0 = I$). Then the set $\{W_1, \cdots, W_v\}$ is a mixing set along $\mathcal{F}^{(3)}$.

By Lemma 3.3, we may pass to a subset of A (which we continue to call A) having measure at least half that of the original A and for which $f = 1_A \in AP(Z, Y, G, \mathcal{F}^{(3)})$. There exists a number $c = c\big(\mu(A)\big) > 0$ and a set $B \in \mathcal{B}$ with $\nu(B) > c$ such that $\gamma_y(A) \ge c$ for all $y \in B$. We now pick a number ϵ satisfying $0 < \epsilon < \frac{c}{2}$ and $\epsilon < \frac{1}{4(t+1)}\sqrt{c^{v+1}}$.

Since $f \in AP(Z, Y, G, \mathcal{F}^{(3)})$, there exist functions $g_1, \cdots, g_l \in \mathrm{L}^2(Z, \mathcal{C}, \gamma)$ and set $D \in \mathcal{B}$ with $\nu(D) < \epsilon$ such that for every $\delta > 0$ and every $T(\alpha_1, \cdots, \alpha_m) \in \mathbf{PE}(G, \mathcal{F}^{(3)})$ there exists $\alpha_0 \in \mathcal{F}^{(3)}$ such that for every $(\alpha_1, \cdots, \alpha_m) \in (\mathcal{F}^{(3)})_<^m$ with $\alpha_1 > \alpha_0$ there is a set

$$E = E(\alpha_1, \cdots, \alpha_m) \in \mathcal{B}$$

with $\nu(E) < \delta$ such that whenever $y \in (D \cup E)^c$, there exists $j = j(\alpha_1, \cdots, \alpha_m, y)$ with $1 \le j \le l$ such that

$$\big|\big|T(\alpha_1, \cdots, \alpha_m)f - g_j\big|\big|_y < \epsilon.$$

Let $B' = \big(B \cap D^c\big)$. Then $\nu(B') > \frac{c}{2}$.

By Theorem 2.12, there exist natural numbers $N = N(t, d, l)$ and $w = w(t, d, l)$, and sets of polynomial expressions

$$L = \big\{L_i(\alpha_1, \cdots, \alpha_N)\big\}_{i=1}^w \subset \mathbf{PE}(G, \mathcal{F}^{(3)})$$

and

$$M = \big\{M_i(\alpha_1, \cdots, \alpha_N)\big\}_{i=1}^w \subset \mathbf{PE}_{\le d}(N)$$

having the property that for any $l-$cell partition $L \times M = \bigcup_{i=1}^l C_i$ there exist numbers a, b and q, with $1 \le a, b \le w$ and $1 \le q \le l$, and sets

$$A_i \subset \{1, \cdots, N\}, \ \ 1 \le i \le m,$$

with $A_1 < \cdots < A_m$, such that under the symbolic substitution $\beta_i = \bigcup_{n \in A_i} \alpha_n$, $1 \le i \le m$, we have

$$\Big\{\big(L_a(\alpha_1, \cdots, \alpha_N)R_i(\beta_1, \cdots, \beta_m), \\ M_b(\alpha_1, \cdots, \alpha_N)W_j(\beta_1, \cdots, \beta_m)\big) : \ 0 \le i, j \le t\Big\} \subset C_q.$$

We may assume that $L_1(\alpha_1,\cdots,\alpha_N)=M_1(\alpha_1,\cdots,\alpha_N)=I$.

$(\mathcal{B},\mathcal{F}^{(1)})$ has the IPPSZ property of degree d, and $\nu(B')>0$. Let

$$a_1=a(B',N,w^2,d)>0 \tag{5.3}$$

and let $Q=w^2l2^{Nm}$. We are now finally in a position to make the value of a explicit:

$$a=a(A,m,t,d)=\frac{a_1c^{v+1}}{16Q}.$$

One may check that a doesn't depend on the set $\{S_0,\cdots,S_t\}$. Indeed, a_1 depends on (B',N,w,d), v is at most t, Q depends on (w,l,N,m), c depends on A, B' depends on B and D, N and w depend on (t,d,l), B, D and l depend on A. (Thus reducing a to dependence on (A,m,t,d).) We now pass to a refinement of $\mathcal{F}^{(3)}$ (again we will continue to call it $\mathcal{F}^{(3)}$) having the following properties:

(i) (See Definition 3.1.) For every $(\alpha_1,\cdots,\alpha_N)\in(\mathcal{F}^{(3)})^N_<$ there exists a set $E=E(\alpha_1,\cdots,\alpha_N)\in\mathcal{B}$ with $\nu(E)<\frac{a_1}{2w}$ such that whenever $y\in\big(D\cup E\big)^c$ there exist numbers $j_i=j_i(\alpha_1,\cdots,\alpha_N,y)$ such that

$$\big|\big|L_i(\alpha_1,\cdots,\alpha_N)f-g_{j_i}\big|\big|_y<\epsilon,\ \ 1\le i\le w.$$

(ii) (See Theorem 4.12.) For every $(\alpha_1,\cdots,\alpha_m)\in(\mathcal{F}^{(3)})^m_<$ we have

$$\Big|\Big|E\Big(\prod_{i=0}^{v}W_i(\alpha_1,\cdots,\alpha_m)f\big|Y\Big)-\prod_{i=0}^{v}W_i(\alpha_1,\cdots,\alpha_m)E(f|Y)\Big|\Big|<\frac{c^{v+1}}{2}\sqrt{\frac{a_1}{4Q}}.$$

According to (5.3) there exists an N-tuple $(\alpha_1,\cdots,\alpha_N)\in(\mathcal{F}^{(3)})^N_<$ (which we now fix) satisfying

$$\nu\Big(\bigcap_{i=1}^{w}\bigcap_{n=1}^{w}\big(L_i(\alpha_1,\cdots,\alpha_N)M_n(\alpha_1,\cdots,\alpha_N)\big)^{-1}B'\Big)\ge a_1.$$

Let

$$C=\Big(\bigcap_{i=1}^{w}\bigcap_{n=1}^{w}\big(L_i(\alpha_1,\cdots,\alpha_N)M_n(\alpha_1,\cdots,\alpha_N)\big)^{-1}B'\Big)$$
$$\setminus\bigcup_{j=1}^{w}\big(M_j(\alpha_1,\cdots,\alpha_N)\big)^{-1}E(\alpha_1,\cdots,\alpha_N).$$

Then $C\in\mathcal{B}$, and since $\nu\big(E(\alpha_1,\cdots,\alpha_N)\big)<\frac{a_1}{2w}$, $\nu(C)>\frac{a_1}{2}$. For every $y\in C$ and every i,n satisfying $1\le i,n\le w$ we have $L_i(\alpha_1,\cdots,\alpha_N)M_n(\alpha_1,\cdots,\alpha_N)y\in B'$, so that in particular there exists a number $j=j(i,n,y)$ with $1\le j\le l$ such that

$$\big|\big|L_i(\alpha_1,\cdots,\alpha_N)f-g_j\big|\big|_{M_n(\alpha_1,\cdots,\alpha_N)y}<\epsilon.$$

Moreover (recall that $L_1=I$), $M_n(\alpha_1,\cdots,\alpha_N)y\in\big(B'\setminus E(\alpha_1,\cdots,\alpha_N)\big)$. Finally we note that $C\subset B'\subset B$ (since $M_1(\alpha_1,\cdots,\alpha_N)=I$ as well).

The function $j(i,n,y)$ induces, for every $y\in C$, an l-cell partition $L\times M=\bigcup_{p=1}^{l}C_p$ by the rule $(L_i,M_n)\in C_p$ if $j(i,n,y)=p$. Hence, there exist numbers p,b and q, with $1\le p,b\le w$ and $1\le q\le l$, and sets $A_1,\cdots,A_m\subset\{1,\cdots,N\}$ with

$A_1 < \cdots < A_m$ (depending on y) such that, under the substitution $\beta_i = \bigcup_{n\in A_i} \alpha_n$, $1 \le i \le m$, we have

$$\big\{\big(L_p(\alpha_1,\cdots,\alpha_N)R_i(\beta_1,\cdots,\beta_m), M_b(\alpha_1,\cdots,\alpha_N)W_n(\beta_1,\cdots,\beta_m)\big) : \\ 0 \le n \le v,\ 0 \le i \le t\big\} \subset C_q \subset L \times M.$$

This implies that for $0 \le n \le v$, $0 \le i \le t$,

$$\Big|\Big|L_p(\alpha_1,\cdots,\alpha_N)R_i(\beta_1,\cdots,\beta_m)f - g_q\Big|\Big|_{M_b(\alpha_1,\cdots,\alpha_N)W_n(\beta_1,\cdots,\beta_m)y} < \epsilon. \tag{5.4}$$

There are not more than $w^2 l 2^{Nm} = Q$ choices for $p, b, q, A_1, \cdots, A_m$. It follows that there exists a set $C' \subset C$ with $\nu(C') \ge \frac{a_1}{2Q}$, and fixed $p, b, q, A_1, \cdots, A_m$ such that (5.4) holds for all $y \in C'$.

Having fixed p and b, it will be convenient for us to adopt the following notation: For $y \in Y$, write $\tilde{y} = L_p(\alpha_1,\cdots,\alpha_N)M_b(\alpha_1,\cdots,\alpha_N)y$. Now (5.4) can be written as follows:

$$\Big|\Big|R_i(\beta_1,\cdots,\beta_m)W_n(\beta_1,\cdots,\beta_m)f - W_n(\beta_1,\cdots,\beta_m)\big(L_p(\alpha_1,\cdots,\alpha_N)^{-1}g_q\big)\Big|\Big|_{\tilde{y}} < \epsilon. \tag{5.5}$$

This holds for $0 \le n \le v$, $0 \le i \le t$ and $y \in C'$. Taking $i = 0$ in (5.5) gives

$$\Big|\Big|W_n(\beta_1,\cdots,\beta_m)f - W_n(\beta_1,\cdots,\beta_m)\big(L_p(\alpha_1,\cdots,\alpha_N)^{-1}g_q\big)\Big|\Big|_{\tilde{y}} < \epsilon \tag{5.6}$$

for $0 \le n \le v$ and $y \in C'$. Combining (5.5) and (5.6) we obtain

$$\Big|\Big|R_i(\beta_1,\cdots,\beta_m)W_n(\beta_1,\cdots,\beta_m)f - W_n(\beta_1,\cdots,\beta_m)f\Big|\Big|_{\tilde{y}} < 2\epsilon,$$

which holds for $0 \le n \le v$, $0 \le i \le t$ and $y \in C'$. Since f is $\{0,1\}$-valued and since for all i with $0 \le i \le t$ there exists n with $0 \le n \le v$ such that $W_i = W_n$, we can observe that

$$\begin{aligned} &\Big|\Big|\prod_{i=0}^{t} S_i(\beta_1,\cdots,\beta_m)f - \prod_{n=0}^{v} W_n(\beta_1,\cdots,\beta_m)f\Big|\Big|_{\tilde{y}} \\ =&\Big|\Big|\prod_{i=0}^{t} R_i(\beta_1,\cdots,\beta_m)W_i(\beta_1,\cdots,\beta_m)f - \prod_{i=0}^{t} W_i(\beta_1,\cdots,\beta_m)f\Big|\Big|_{\tilde{y}} \\ <&2(t+1)\epsilon,\ y \in C'. \end{aligned} \tag{5.7}$$

Recall that $C' \subset C$, and for $y \in C$ we have $L_i(\alpha_1,\cdots,\alpha_N)M_n(\alpha_1,\cdots,\alpha_N)y \in B' \subset B$ for all $L_i \in L$ and $M_n \in M$. In particular, since

$$M_b(\alpha_1,\cdots,\alpha_N)W_i(\beta_1,\cdots,\beta_m) \in M,\ \ 0 \le i \le v,$$

we have, for $0 \le i \le v$,

$$W_i(\beta_1,\cdots,\beta_m)\tilde{y} = L_p(\alpha_1,\cdots,\alpha_N)M_b(\alpha_1,\cdots,\alpha_N)W_i(\beta_1,\cdots,\beta_m)y \in B.$$

Recall as well that on B, $E(f|Y) \ge c$. Because of this, for all $y \in C'$ we have

$$\Big|\prod_{i=0}^{v} W_i(\beta_1,\cdots,\beta_m)E(f|Y)(\tilde{y})\Big| \ge c^{v+1}.$$

On the other hand, from (ii) (see above) we can see that for all y outside of a set of measure at most $\frac{a_1}{4Q}$,

$$\Big|E\Big(\prod_{i=0}^{v} W_i(\beta_1,\cdots,\beta_m)f|Y\Big)(\tilde{y}) - \prod_{i=0}^{v} W_i(\beta_1,\cdots,\beta_m)E(f|Y)(\tilde{y})\Big| < \frac{c^{v+1}}{2}.$$

But $\nu(C') \geq \frac{a_1}{2Q}$. Hence there exists $C'' \in \mathcal{B}$ with $C'' \subset C'$ and $\nu(C'') \geq \frac{a_1}{4Q}$ such that for all $y \in C''$,

$$E\Big(\prod_{i=0}^{v} W_i(\beta_1,\cdots,\beta_m)f|Y\Big)(\tilde{y}) > \frac{c^{v+1}}{2}. \tag{5.8}$$

Here is a general fact: for $\{0,1\}$-valued functions f and g, $|f-g| = |f-g|^2$, so that $(\int f - \int g) \leq \int |f-g| = \int |f-g|^2 = \sqrt{||f-g||}$. Therefore by (5.7) we see that for all $y \in C'$ (in particular, for all $y \in C''$),

$$E\Big(\prod_{i=0}^{t} S_i(\beta_1,\cdots,\beta_m)f|Y\Big)(\tilde{y}) \geq E\Big(\prod_{n=0}^{v} W_n(\beta_1,\cdots,\beta_m)f|Y\Big)(\tilde{y}) - 4(t+1)^2\epsilon^2.$$

Together with (5.8) this gives (recall that $\epsilon < \frac{1}{4(t+1)}\sqrt{c^{v+1}}$), for all $y \in C''$,

$$E\Big(\prod_{i=0}^{t} S_i(\beta_1,\cdots,\beta_m)f|Y\Big)(\tilde{y}) > \frac{c^{v+1}}{4}.$$

This, since $\nu(C'') \geq \frac{a_1}{4Q}$, is sufficient to guarantee that

$$\gamma\Big(\bigcap_{i=0}^{t} S_i(\beta_1,\cdots,\beta_m)^{-1}A\Big) \geq \frac{a_1 c^{v+1}}{16Q} = a.$$

□

Remark 5.4 Our work so far establishes Theorem 1.3 for Lebesgue spaces. The theorem holds for spaces which are not Lebesgue as well. Let $(X,\mathcal{A},\mu,\Omega)$ be a measure preserving system where the underlying space is not Lebesgue. As we have stated, it suffices to assume that the system is separable. Let us identify two measurable sets A and B if $\mu(A\triangle B)=0$. The *measure algebra* of the system $(X,\mathcal{A},\mu,\Omega)$ is the set of equivalence classes represented in $\mathcal{A}$ under this identification. It is well known that any probability measure preserving system is equivalent (via an isomorphism of the measure algebras which commutes with the measure preserving actions) to a system $(X',\mathcal{A}',\mu',\Omega')$ whose underlying space is Lebesgue. (See, for example, Chapter 15 of [R] or Theorem 5.15 of [F2].) Since validity of the conclusion of Theorem 1.3 is clearly preserved by measure algebra isomorphism (even when that isomorphism does not come from a pointwise map), its truth for $(X,\mathcal{A},\mu,\Omega)$ follows from its truth for $(X',\mathcal{A}',\mu',\Omega')$.

CHAPTER 6

MEASURE THEORETIC APPLICATIONS

This section is devoted to applications of (special cases of) our main theorem to properties of multiple recurrence return times. The first thing we shall do is show, as promised in Section 1, how one reduces the case of polynomials in $\mathbf{Q}[x_1, \cdots, x_k]$ which take on integer values on the integers to the case of polynomials in $\mathbf{Z}[x_1, \cdots, x_k]$. For example, consider Theorem 0.9:

Theorem 0.9 Suppose that we have r commuting measure preserving transformations $T_1, \cdots, T_r$ of a probability space $(X, \mathcal{B}, \mu)$. Suppose $k, t \in \mathbf{N}$, and $p_{i,j}(x_1, \cdots, x_k) \in \mathbf{Q}[x_1, \cdots, x_k]$ with $p_{i,j}(\mathbf{Z}^k) \subset \mathbf{Z}$ and $p_{i,j}(0, \cdots, 0) = 0$, $1 \le i \le r$, $1 \le j \le t$. Then for every $A \in \mathcal{B}$ with $\mu(A) > 0$ the set

$$R_A = \Big\{(n_1, \cdots, n_k) \in \mathbf{Z}^k : \mu\Big(\bigcap_{j=1}^{t}\Big(\prod_{i=1}^{r} T_i^{p_{i,j}(n_1,\cdots,n_k)}\Big)^{-1} A\Big) > 0\Big\}$$

is an IP*-set in $\mathbf{Z}^k$.

Let us show how to derive this result from Theorem 1.3. Let N be a natural number which is a multiple of every denominator of every coefficient of every polynomial $p_{i,j}$, $1 \le i \le r$, $1 \le j \le t$. Then for all i and r with $1 \le i \le r$ and $1 \le j \le t$ the polynomial

$$q_{i,j}(x_1, \cdots, x_k) = p_{i,j}(Nx_1, \cdots, Nx_k)$$

is in $\mathbf{Z}[x_1, \cdots, x_k]$. Suppose $(\mathbf{n}_\alpha)_{\alpha\in\mathcal{F}}$ is an IP-set in $\mathbf{Z}^k$. Write $\mathbf{n}_\alpha = (n_\alpha^{(1)}, \cdots, n_\alpha^{(k)})$. Then $(n_\alpha^{(i)})_{\alpha\in\mathcal{F}}$ is an IP-set in $\mathbf{Z}$ for all i, $1 \le i \le k$. Let $C_i = \{\alpha \in \mathcal{F} : n_\alpha^{(1)} = i \mod i\}$, $0 \le i < N$. One may easily check elementarily that C_0 contains an IP-ring. Alternatively, $\mathcal{F} = \bigcup_{i=0}^{N-1} C_i$, so by Theorem 2.13 one of the cells of this partition must contain an IP-ring. Furthermore, it is easily seen that C_0 is the only cell which can possibly contain an IP-ring. Therefore C_0 contains an IP-ring, say $\mathcal{F}^{(1)}$. $\mathcal{F}^{(1)}$ has the property that the $(k_\alpha^{(1)})_{\alpha\in\mathcal{F}^{(1)}} \subset \mathbf{Z}$, where $k_\alpha^{(1)} = \frac{1}{N} n_\alpha^{(1)}$. By a repetition of this argument in the remaining indices we may require of $\mathcal{F}^{(1)}$ that the IP-sets $(k_\alpha^{(i)})_{\alpha\in\mathcal{F}^{(1)}}$, $1 \le i \le k$, are all contained in $\mathbf{Z}$, where $k_\alpha^{(i)} = \frac{1}{N} n_\alpha^{(i)}$.

Let d be the maximum degree of the polynomials $p_{i,j}$ and let $\mathbf{PE}_{\le d}(1)$ be as in Definition 1.1 (that is, for these fixed values of k, r, and d). Put

$$S_j(\alpha) = \Big(\prod_{i=1}^{r} T_i^{q_{i,j}(k_\alpha^{(1)},\cdots,k_\alpha^{(k)})}\Big), \ 1 \le j \le t.$$

Then $S_j \in \mathbf{PE}_{\le d}(1)$, $1 \le j \le t$.

According to the case $m = 1$ of Theorem 1.3, there exists $\alpha \in \mathcal{F}^{(1)}$ such that

$$\mu\Big(\bigcap_{j=1}^{t} S_j(\alpha)^{-1}A\Big) > 0.$$

In other words,

$$\mu\Big(\bigcap_{j=1}^{t}\Big(\prod_{i=1}^{r} T_i^{q_{i,j}(k_\alpha^{(1)},\cdots,k_\alpha^{(k)})}\Big)^{-1}A\Big) > 0,$$

which is the same as

$$\mu\Big(\bigcap_{j=1}^{t}\Big(\prod_{i=1}^{r} T_i^{p_{i,j}(n_\alpha^{(1)},\cdots,n_\alpha^{(k)})}\Big)^{-1}A\Big) > 0.$$

Hence $\mathbf{n}_\alpha \in R_A$. Since the IP-set $(\mathbf{n}_\alpha)_{\alpha\in\mathcal{F}^{(1)}}$ was arbitary, we have shown that R_A is an IP*-set.

□

The reduction to polynomials in $\mathbf{Z}[x_1, \cdots, x_k]$ may be accomplished in an identical fashion in all of our remaining applications. Henceforth, we will not carry this reduction out explicitly, but rather simply assume without loss of generality that we are dealing with polynomials with integer coefficients.

The following definition is made for the purpose of showing that the set R_A of Theorem 0.9 (for example) has in fact a stronger combinatorial structure than does the typical IP*-set.

Definition 6.1 Suppose $k \in \mathbf{N}$. A set $S \subset \mathbf{Z}^k$ is said to be an *enhanced IP*-set*, or an *E-IP*-set*, if for any $m \in \mathbf{N}$ and any m IP-sets $(n_\alpha^{(i)})_{\alpha\in\mathcal{F}_\emptyset}$, $1 \le i \le m$ there exists an IP-ring $\mathcal{F}^{(1)}$ such that

$$\{n_{\alpha_1}^{(1)} + n_{\alpha_2}^{(2)} + \cdots + n_{\alpha_m}^{(m)} : (\alpha_1, \cdots, \alpha_m) \in (\mathcal{F}_\emptyset^{(1)})^m\} \subset S.$$

$S \subset \mathbf{Z}^k$ is said to be a *polynomially enhanced IP*-set*, or a *PE-IP*-set*, if for any $m \in \mathbf{N}$, any m IP-sets $(n_\alpha^{(i)})_{\alpha\in\mathcal{F}_\emptyset} \subset \mathbf{Z}^k$, $1 \le i \le m$, and any polynomial mapping $P : \mathbf{Z}^m \to \mathbf{Z}^k$ having integer coefficients and satisfying $P(0) = 0$ there exists an IP-ring $\mathcal{F}^{(1)}$ such that (recall that $n_\emptyset^{(i)} = 0$)

$$\{P(n_{\alpha_1}^{(1)}, n_{\alpha_2}^{(2)}, \cdots, n_{\alpha_m}^{(m)}) : (\alpha_1, \cdots, \alpha_m) \in (\mathcal{F}_\emptyset^{(1)})^m\} \subset S.$$

We call a set $S \subset \mathbf{Z}^k$ an *IP$^*_+$-set* if some shift of S is an IP*-set. That is, if for some $u \in \mathbf{Z}^k$, $u + S = \{u + s : s \in S\}$ is an IP*-set. Similarly, if $S \subset \mathbf{Z}^k$ and $u + S = \{u + s : s \in S\}$ is an E-IP*- (respectively PE-IP*-) set for some $u \in \mathbf{Z}^k$ then S is said to be an *E-IP$^*_+$-set* (respectively *PE-IP$^*_+$-set*).

IP$^*_+$-sets are clearly syndetic, but the IP$^*_+$ property is stronger than syndeticity, as we show in Example 7.7.

It is a consequence of Hindman's theorem that the intersection of any two IP*-sets is an IP*-set. E-IP*-sets and PE-IP*-sets have the finite intersection property as well, although this is a very elementary observation which does not require Hindman's theorem. Examples will be given in Section 7 which show that the notions of IP*, E-IP* and PE-IP* are distinct.

Notice that in the previous definition, we consider m-tuples $(\alpha_1, \cdots, \alpha_m) \in (\mathcal{F}_\emptyset)^m$ which need *not* be in $(\mathcal{F}_\emptyset)^m_<$. Our motivation for doing this is the following lemma and its corollary, Theorem 6.3. Notice that Lemma 6.2 is, in one sense, a strengthening of Theorem 1.3. In its formulation we need to expand the definition of a polynomial expression to take for its argument any m-tuple $(\alpha_1, \cdots, \alpha_m) \in (\mathcal{F}_\emptyset)^m$. Recall (Definition 1.1) that a polynomial expression in $\mathbf{PE}(m)$ is of the form

$$T(\alpha_1, \cdots, \alpha_m) = \prod_{i=1}^{r} T_i^{p_i\left((n^{(b)}_{\alpha_j})_{1 \le b \le k,\ 1 \le j \le m}\right)}, \quad (\alpha_1, \cdots, \alpha_m) \in (\mathcal{F}_\emptyset)^m_<$$

One need merely note that the above form of $T(\alpha_1, \cdots, \alpha_m)$ makes just as much sense (as a function onto measure preserving transformations) for all $(\alpha_1, \cdots, \alpha_m) \in (\mathcal{F}_\emptyset)^m$. Indeed, the reason we did not allow for this possibility sooner is that when writing such expressions as $S(\alpha_1, \alpha_2) = T(\alpha_1 \cup \alpha_2)$, we wanted to ensure that $S(\alpha_1, \alpha_2) \in \mathbf{PE}(2)$. This required that $n^{(i)}_{\alpha \cup \beta} = n^{(i)}_\alpha + n^{(i)}_\beta$, $1 \le i \le k$.

Lemma 6.2 Let $k, r \in \mathbf{N}$ and let $\mathbf{PE}_{\le d}(m)$ be as in Definition 1.1. Suppose that $(n^{(i)}_\alpha)_{\alpha \in \mathcal{F}}$ are IP-sets, $1 \le i \le k$, and let $(X, \mathcal{A}, \mu, \Omega)$ be a measure preserving system, where Ω is generated by commuting invertible transformations $T_1, \cdots, T_r$. If $S_0, \cdots, S_t \in \mathbf{PE}_{\le d}(m)$ and $\mathcal{F}^{(1)}$ is any IP-ring then there exists an IP-ring $\mathcal{F}^{(2)} \subset \mathcal{F}^{(1)}$ such that for all $(\alpha_1, \cdots, \alpha_m) \in (\mathcal{F}^{(2)}_\emptyset)^m$ we have

$$\mu\Big(\bigcap_{i=0}^{t} S_i(\alpha_1, \cdots, \alpha_m)^{-1} A\Big) > 0.$$

Proof. We will inductively construct a sequence $(\beta_i)_{i \in \mathbf{N}} \subset \mathcal{F}^{(1)}$ with $\beta_1 < \beta_2 < \cdots$ such that the IP-ring $\mathcal{F}^{(2)} = FU\{(\beta_i)_{i \in \mathbf{N}}\}$ has the properties we require. Specifically, we will construct a sequence $(\beta_i)_{i \in \mathbf{N}}$ and a sequence of measurable sets $A = A_0 \supset A_1 \supset A_2 \supset \cdots$ with $\mu(A_i) > 0$ for all $i \in \mathbf{N}$ having the property that for all $i \in \mathbf{N}$, every $x \in A_i$, and every choice of $\gamma_1, \gamma_2, \cdots, \gamma_m \in FU_\emptyset\{\beta_1, \cdots, \beta_i\}$ we have

$$S_j(\gamma_1, \cdots, \gamma_m)x \in A, \quad 0 \le j \le t.$$

Let $A_0 = A$ and let $\mathcal{T}_1$ be the family of all polynomial expressions of the form

$$T(\alpha) = S_j\big(\chi_1(\alpha), \cdots, \chi_m(\alpha)\big),$$

where $1 \le j \le t$ and for $1 \le i \le m$, either $\chi_i(\alpha) = \alpha$ or $\chi_i(\alpha) = \emptyset$. By Theorem 1.3 there exists $\beta_1 \in \mathcal{F}^{(1)}$ such that the set

$$A_1 = \bigcap_{T \in \mathcal{T}_1} T(\beta_1)^{-1} A$$

satisfies $\mu(A_1) > 0$. One may easily check that for all $x \in A_1$ and every choice of $\gamma_1, \gamma_2, \cdots, \gamma_m \in FU_\emptyset\{\beta_1\} = \{\emptyset, \beta_1\}$ we have $S_j(\gamma_1, \cdots, \gamma_m)x \in A$, $0 \le j \le t$.

Suppose now that $\beta_1, \beta_2, \cdots, \beta_i$ and $A_1, \cdots, A_i$ have been chosen. Let $\mathcal{T}_{i+1}$ be the family of all polynomial expressions of the form

$$T(\alpha) = S_j\big(\alpha_1 \cup \chi_1(\alpha), \cdots, \alpha_m \cup \chi_m(\alpha)\big) S_j(\alpha_1, \cdots, \alpha_m)^{-1},$$

where $0 \leq j \leq t$, $\alpha_1, \cdots, \alpha_m \in FU_\emptyset\{\beta_1, \cdots, \beta_i\}$ and for $1 \leq i \leq m$, either $\chi_i(\alpha) = \alpha$ or $\chi_i(\alpha) = \emptyset$. By Theorem 1.3 there exists $\beta_{i+1} \in \mathcal{F}^{(1)}$ with $\beta_{i+1} > \beta_i$ such that the set

$$A_{i+1} = \bigcap_{T \in \mathcal{T}_{i+1}} T(\beta_{i+1})^{-1} A_i$$

satisfies $\mu(A_{i+1}) > 0$. We claim that for all $x \in A_{i+1}$ and every choice of $\gamma_1, \gamma_2, \cdots, \gamma_m \in FU_\emptyset\{\beta_1, \beta_2, \cdots, \beta_{i+1}\}$ one has

$$S_j(\gamma_1, \cdots, \gamma_m)x \in A, \;\; 1 \leq j \leq t.$$

Indeed, let $\alpha_n = \gamma_n \setminus \beta_{i+1}$, $1 \leq n \leq m$. Then

$$S_j(\gamma_1, \cdots, \gamma_m) S_j(\alpha_1, \cdots, \alpha_m)^{-1} x \in A_i \subset S_j(\alpha_1, \cdots, \alpha_m)^{-1} A, \; 1 \leq j \leq t.$$

In particular $S_j(\gamma_1, \cdots, \gamma_m)x \in A$, $1 \leq j \leq t$, as required. Therefore,

$$\mu\Big(\bigcap_{i=0}^{t} S_i(\gamma_1, \cdots, \gamma_m)^{-1} A\Big) > 0$$

for every choice of $\gamma_1, \gamma_2, \cdots, \gamma_m \in FU_\emptyset\{\beta_1, \beta_2, \cdots, \beta_{i+1}\}$. Continue choosing the β_i's in this fashion. Finally, let $\mathcal{F}^{(2)} = FU\{\beta_1, \beta_2, \cdots\}$.

□

We now present our main theorem pertaining to largeness of sets of polynomial multiple recurrence return times.

Theorem 6.3 Suppose that we have r commuting invertible measure preserving transformations $T_1, \cdots, T_r$ of a probability space $(X, \mathcal{B}, \mu)$. Let $k, t \in \mathbf{N}$, and assume that $p_{i,j}(x_1, \cdots, x_k) \in \mathbf{Q}[x_1, \cdots, x_k]$ with $p_{i,j}(\mathbf{Z}^k) \subset \mathbf{Z}$ and $p_{i,j}(0, \cdots, 0) = 0$, $1 \leq i \leq r$, $1 \leq j \leq t$. Then for every $A \in \mathcal{B}$ with $\mu(A) > 0$ the set

$$R_A = \Big\{(n_1, \cdots, n_k) \in \mathbf{Z}^k : \mu\Big(\bigcap_{j=1}^{t} \Big(\prod_{i=1}^{r} T_i^{p_{i,j}(n_1, \cdots, n_k)}\Big)^{-1} A\Big) > 0\Big\}$$

is a PE-IP*-set in $\mathbf{Z}^k$.

Proof. We may assume without loss of generality that

$$p_{i,j}(x_1, \cdots, x_k) \in \mathbf{Z}[x_1, \cdots, x_k], \;\; 1 \leq i \leq r, \; 1 \leq j \leq t.$$

Suppose that $m \in \mathbf{N}$ and let $(n_\alpha^{(1)})_{\alpha \in \mathcal{F}}, \cdots, (n_\alpha^{(m)})_{\alpha \in \mathcal{F}}$ be IP-sets in $\mathbf{Z}$. Suppose that $P : \mathbf{Z}^m \to \mathbf{Z}^k$ is a polynomial mapping with integer coefficients and with $P(0) = 0$. We have to find an IP-ring $\mathcal{F}^{(1)}$ such that $P(n_{\alpha_1}^{(1)}, \cdots, n_{\alpha_m}^{(m)}) \in R_A$, or equivalently

$$\mu\Big(\bigcap_{j=1}^{t} \Big(\prod_{i=1}^{r} T_i^{p_{i,j}(P(n_{\alpha_1}^{(1)}, \cdots, n_{\alpha_m}^{(m)}))}\Big)^{-1} A\Big) > 0,$$

for all m-tuples $(\alpha_1, \cdots, \alpha_m) \in (\mathcal{F}_\emptyset^{(1)})^m$.

Let $q_{i,j} = p_{i,j} \circ P$, $1 \leq i \leq i$, $1 \leq j \leq t$, and put

$$S_j(\alpha_1, \cdots, \alpha_m) = \prod_{i=1}^{r} T_i^{q_{i,j}(n_{\alpha_1}^{(1)}, \cdots, n_{\alpha_m}^{(m)})}.$$

Then $S_j \in \mathbf{PE}(m)$, $1 \leq j \leq t$, so by Lemma 6.2 there exists an IP ring $\mathcal{F}^{(1)}$ such that for all m-tuples $(\alpha_1, \cdots, \alpha_m) \in (\mathcal{F}^{(1)})^m$ we have

$$\mu\Big(\bigcap_{j=1}^{t} \Big(\prod_{i=1}^{r} T_i^{p_{i,j}(P(n_{\alpha_1}^{(1)}, \cdots, n_{\alpha_m}^{(m)}))}\Big)^{-1} A\Big) = \mu\Big(\bigcap_{j=1}^{r} S_j(\alpha_1, \cdots, \alpha_m)^{-1} A\Big) > 0.$$

□

According to Theorem 6.3, the set

$$R_A = \Big\{(n_1, \cdots, n_k) \in \mathbf{Z}^k : \mu\Big(\bigcap_{j=1}^{t} \Big(\prod_{i=1}^{r} T_i^{p_{i,j}(n_1, \cdots, n_k)}\Big)^{-1} A\Big) > 0\Big\}$$

is a PE-IP*-set. We now would like to investigate under what conditions one may say something about the set

$$R_{A_0, A_1, \cdots, A_t} = \Big\{(n_1, \cdots, n_k) \in \mathbf{Z}^k : \mu\Big(A_0 \cap \bigcap_{j=1}^{t} \Big(\prod_{i=1}^{r} T_i^{p_{i,j}(n_1, \cdots, n_k)}\Big)^{-1} A_j\Big) > 0\Big\},$$

where $A_0, A_1, \cdots, A_t \in \mathcal{B}$ are any positive measure sets.

In the case $t = 1$, $r = 1$, $k = 1$, $p_{1,1}(n) = n$, one may show that the set R_{A_0, A_1} is in fact a shift of an IP*-set provided that $T = T_1$ is ergodic.

In fact, when T is ergodic, for any $\epsilon < \mu(A_0)\mu(A_1)$ the set $R = \{n : \mu(A_0 \cap T^{-n} A_1) > \epsilon^2\}$, which is a subset of R_{A_0, A_1}, is an IP_+^*-set. To see this, consider that by ergodicity there exists $k \in \mathbf{Z}$ such that $\mu(A_0 \cap T^{-k} A_1) > \epsilon$. Let now $B = (A_0 \cap T^{-k} A_1)$. It is well known that (since $\epsilon^2 < \mu(B)^2$) the set $S = \{n : \mu(B \cap T^{-n} B) > \epsilon^2\}$ is an IP*-set (see, for example, [B2]). However, $n \in S$ implies $n + k \in R$, so that $S \subset R - k$ and $R - k$ is an IP*-set.

The situation is different in the case of polynomial powers, when ergodicity of T is no longer sufficient to guarantee, for example, even a single non-zero number $n \in \mathbf{Z}$ for which $\mu(A_0 \cap T^{-n^2} A_1) > 0$. (Consider the system $X = \{0, 1, 2\}$, with $\mu(\{i\}) = \frac{1}{3}$, $i = 0, 1, 2$ and $Ti = i + 1 \pmod 3$. Then $\mu(\{0\} \cap T^{-n^2}\{2\}) = 0$ for all $n \in \mathbf{Z}$, yet T is ergodic.) In the case of *total ergodicity* (T is totally ergodic if T^k is ergodic for all $k \in \mathbf{N}$), one may show (we leave it as an exercise) that for any $\epsilon < \mu(A_0)\mu(A_1)$ the set

$$U = \{n : \mu(A_0 \cap T^{-n^2} A_1) > \epsilon^2\}$$

is an IP_+^*-set.

Question 1. Can a larger lower bound than $\mu(A_0)^2\mu(A_1)^2$ be found for the size of the intersection in either of these cases? (In the event that $A_0 = A_1$, $\mu(A_0)\mu(A_1)$ works.)

Total ergodicity of T is insufficient to guarantee even a single non-zero member of $R_{A_0, A_1, \cdots, A_t}$, however, in the situation of multiple recurrence (i.e. $t \geq 2$). For example, if α is any rational number, then setting $X = [0, 1)$ and $Tx = x + \alpha \pmod 1$, T is a totally ergodic transformation of X, but there exist positive measure sets A_0, A_1, and A_2 such that $\mu(A_0 \cap T^{-n} A_1 \cap T^{-2n} A_2) = 0$ for all $n \in \mathbf{Z}$. Indeed, one may let $A_0 = A_1 = [0, \frac{1}{4})$ and $A_2 = [\frac{1}{2}, \frac{3}{4})$.

A correct condition for largeness of the set $R_{A_0,\cdots,A_r}$ turns out to be that of *total weak mixing*. Recall that a measure preserving $\mathbf{Z}^r$-action $\{T_\mathbf{n}\}_{\mathbf{n}\in\mathbf{Z}^r}$ on a probability measure space $(X, \mathcal{B}, \mu)$ is called totally weakly mixing if for every non-zero $\mathbf{n} \in \mathbf{Z}^r$ $T_\mathbf{n}$ is a weakly mixing transformation, i.e. if for every non-constant $f \in L^2(X, \mathcal{B}, \mu)$ the orbit $\{T_{k\mathbf{n}} f : k \in \mathbf{Z}\}$ fails to be precompact in $L^2(X, \mathcal{B}, \mu)$.

We remark that the total weak mixing condition for a measure preserving $\mathbf{Z}^r$-action differs (when $r \geq 2$) from the mere weak mixing notion. ($\{T_\mathbf{n}\}_{\mathbf{n}\in\mathbf{Z}^r}$ is weakly mixing if for every non-constant $f \in L^2(X, \mathcal{B}, \mu)$ the orbit $\{T_\mathbf{n} f : \mathbf{n} \in \mathbf{Z}^r\}$ fails to be precompact in $L^2(X, \mathcal{B}, \mu)$.) We will illustrate the disparity with an example. Let Y be a probability space and $T : Y \to Y$ an invertible weakly mixing transformation. For $i \in \mathbf{N}$ put $Y_i = Y$ and let $X = \prod_{i=1}^\infty Y_i$ with the product measure. Let $\{(a_i, b_i)\}_{i\in\mathbf{N}}$ be an enumeration of all ordered pairs in $\mathbf{Z}^2$. Define measure preserving transformations S, U on X by $S(y_1, y_2, y_3, \cdots) = (T^{a_1} y_1, T^{a_2} y_2, T^{a_3} y_3, \cdots)$ and $U(y_1, y_2, y_3, \cdots) = (T^{b_1} y_1, T^{b_2} y_2, T^{b_3} y_3, \cdots)$. The measure preserving $\mathbf{Z}^2$-action $\{S^n U^m : n, m \in \mathbf{Z}\}$ is clearly weak mixing on X, however its component transformations fail to be ergodic. (Let $n, m \in \mathbf{Z}$. For some $i \in \mathbf{N}$ we have $(a_i, b_i) = (m, -n)$. In the ith coordinate, $S^n U^m$ is $(T^m)^n (U^{-n})^m = I$, hence $S^n U^m$ is not ergodic.) In particular, this $\mathbf{Z}^2$-action is not totally weakly mixing.

Theorem 6.4 Suppose that $(X, \mathcal{B}, \mu)$ is a probability space and $\{T_\mathbf{n}\}_{\mathbf{n}\in\mathbf{Z}^r}$ is a totally weakly mixing measure preserving $\mathbf{Z}^r$-action generated by $T_1, \cdots, T_r$. Suppose that $t \in \mathbf{N}$ and $p_{i,j}(x_1, \cdots, x_k) \in \mathbf{Q}[x_1, \cdots, x_k]$ with $p_{i,j}(\mathbf{Z}^k) \subset \mathbf{Z}$, $1 \leq i \leq r$, $1 \leq j \leq t$ such that for any $1 \leq j_1 \neq j_2 \leq t$, the functions

$$(l_1, \cdots, l_k) \to (p_{1,j_1}(l_1, \cdots, l_k), p_{2,j_1}(l_1, \cdots, l_k), \cdots p_{r,j_1}(l_1, \cdots, l_k))$$

and

$$\begin{aligned}&(l_1, \cdots, l_k)\\ \to &(p_{1,j_1}(l_1, \cdots, l_k) - p_{1,j_2}(l_1, \cdots, l_k), \cdots, (p_{r,j_1}(l_1, \cdots, l_k) - p_{r,j_2}(l_1, \cdots, l_k)\end{aligned}$$

are not constant. Suppose that $A_0, A_1, A_2, \cdots, A_t \in \mathcal{B}$ with $\mu(A_i) > 0$, $0 \leq i \leq t$. Then the set

$$R_{A_0,A_1,\cdots,A_t} = \left\{(n_1, \cdots, n_k) \in \mathbf{Z}^k : \mu\left(A_0 \cap \bigcap_{j=1}^t \left(\prod_{i=1}^r T_i^{p_{i,j}(n_1,\cdots,n_k)}\right)^{-1} A_i\right) > 0\right\} \tag{6.1}$$

is a PE-IP$^*_+$-set.

Proof. Again, without loss of generality we may assume that all of the polynomials $p_{i,j}$ lie in $\mathbf{Z}[x_1, \cdots, x_k]$. There exists $(l_1, \cdots, l_k) \in \mathbf{Z}^k$ such that

$$\mu\left(A_0 \cap \bigcap_{j=1}^t \left(\prod_{i=1}^r T_i^{p_{i,j}(l_1,\cdots,l_k)}\right)^{-1} A_i\right) > 0. \tag{6.2}$$

(See Appendix, Theorem A1.)

Let

$$B = A_0 \cap \bigcap_{j=1}^{t} \Big(\prod_{i=1}^{r} T_i^{p_{i,j}(l_1,\cdots,l_k)} \Big)^{-1} A_i.$$

Then $\mu(B) > 0$. Letting

$$q_{i,j}(n_1,\cdots,n_k) = p_{i,j}(n_1 + l_1,\cdots,n_k + l_k) - p_{i,j}(l_1,\cdots,l_k)$$

we have $q_{i,j}(0) = 0$, $1 \leq i \leq r$, $1 \leq j \leq t$. Therefore, by Theorem 6.3 the set

$$R = \Big\{ (n_1,\cdots,n_k) \in \mathbf{Z}^k : \mu\Big(\bigcap_{j=1}^{t} \Big(\prod_{i=1}^{r} T_i^{q_{i,j}(n_1,\cdots,n_k)} \Big)^{-1} B \Big) > 0 \Big\}$$

is a PE-IP*-set. One now need only check that $(n_1,\cdots,n_k) \in R$ implies

$$(n_1 + l_1,\cdots,n_k + l_k) \in R_{A_0,A_1,\cdots,A_r}.$$

Therefore $R \subset R_{A_0,A_1,\cdots,A_r} - (l_1,\cdots,l_k)$ and $R_{A_0,A_1,\cdots,A_r}$ is a PE-IP$^*_+$-set.

□

It is a simple corollary of Theorem 2.13 (Hindman's Theorem) that the intersection of two IP*-sets in $\mathbf{Z}^k$ is again an IP*-set. As mentioned earlier, the intersection of two E-IP*-sets is an E-IP*-set. IP$^*_+$-sets do not have such a finite intersection property. (For example, consider the even and odd integers. Both are IP$^*_+$-sets (indeed, both are PE-IP$^*_+$-sets). Consider, however, sets of the type $R_{A_0,A_1,\cdots,A_t}$ appearing in Theorem 6.4. One may show (see Appendix, Theorem A1) that they are very large. Indeed, let us introduce the following notion: suppose that $r \in \mathbf{N}$ and $E \subset \mathbf{Z}^r$ is a set. The *lower Banach density* of E may be defined to be the number $d_*(E) = 1 - d^*(E^c)$, where d^* is upper Banach density (see the third paragraph of the introduction for $d^*(E)$). Clearly sets of Banach lower density 1 are very "big", and a consequence of Theorem A1 is that sets of the type $R_{A_0,A_1,\cdots,A_t}$ appearing in Theorem 6.4 have Banach lower density 1. It follows easily that the intersection in $\mathbf{Z}^k$ of any finite collection of sets having the form (6.1) would again be of uniform lower density 1.

We point out, however, that sets of Banach lower density 1 need not be IP*-sets, nor even IP$^*_+$-sets . (See Example 7.8 for a set of Banach lower density 1 which is not a IP$^*_+$-set . As for a set of Banach lower density 1 which fails to be IP*, if we let $n_\alpha = \sum_{i\in\alpha} 3^i$, $\alpha \in \mathcal{F}$, then $(n_\alpha)_{\alpha\in\mathcal{F}}$ is an IP-set of Banach upper density 0, so that its complement is a set of Banach lower density 1 with fails to be IP*.) The content of the next theorem is that, as a matter of fact, the intersection in $\mathbf{Z}^k$ of any finite collection of sets having the form (6.1) is not merely of Banach lower density 1, but PE-IP$^*_+$ as well.

Theorem 6.5 Suppose that $S \in \mathbf{N}$ and for each s, $1 \leq s \leq S$, the following are given. $(X^{(s)}, \mathcal{B}^{(s)}, \mu^{(s)})$ is a probability space, $r^{(s)} \in \mathbf{N}$, and $\{T_{\mathbf{n}}^{(s)}\}_{\mathbf{n}\in\mathbf{Z}^{r^{(s)}}}$ is a totally weakly mixing measure preserving $\mathbf{Z}^{r^{(s)}}$-action generated by $T_1^{(s)},\cdots,T_{r^{(s)}}^{(s)}$. $t^{(s)} \in \mathbf{N}$ and $p_{i,j}^{(s)}(x_1,\cdots,x_k) \in \mathbf{Q}[x_1,\cdots,x_k]$ with $p_{i,j}^{(s)}(\mathbf{Z}^k) \subset \mathbf{Z}$, $1 \leq i \leq r^{(s)}$, $1 \leq j \leq t^{(s)}$ such that for any $1 \leq j_1 \neq j_2 \leq t^{(s)}$, the functions

$$(l_1,\cdots,l_k) \to (p_{1,j_1}^{(s)}(l_1,\cdots,l_k), p_{2,j_1}^{(s)}(l_1,\cdots,l_k),\cdots p_{r^{(s)},j_1}^{(s)}(l_1,\cdots,l_k))$$

and

$$(l_1, \cdots, l_k)$$
$$\to (p_{1,j_1}^{(s)}(l_1, \cdots, l_k) - p_{1,j_2}^{(s)}(l_1, \cdots, l_k), \cdots, p_{r^{(s)},j_1}^{(s)}(l_1, \cdots, l_k) - p_{r^{(s)},j_2}^{(s)}(l_1, \cdots, l_k))$$

are not constant. Suppose that $A_0^{(s)}, A_1^{(s)}, A_2^{(s)}, \cdots, A_{t^{(s)}}^{(s)} \in \mathcal{B}^{(s)}$ with $\mu(A_i^{(s)}) > 0$, $0 \le i \le t^{(s)}$. Let

$$R_{A_0^{(s)}, A_1^{(s)}, \cdots, A_{t^{(s)}}^{(s)}}$$
$$= \Big\{ (n_1, \cdots, n_k) \in \mathbf{Z}^k : \mu\Big(A_0^{(s)} \cap \bigcap_{j=1}^{t} \Big(\prod_{i=1}^{r} (T_i^{(s)})^{p_{i,j}^{(s)}(n_1, \cdots, n_k)} \Big)^{-1} A_i^{(s)} \Big) > 0 \Big\}.$$

Then $\bigcap_{s=1}^{S} R_{A_0^{(s)}, \cdots, A_{t^{(s)}}^{(s)}}$ is a PE-IP$_+^*$-set.

Proof. Reindexing if necessary, we may assume that $r^{(1)} \ge r^{(2)} \ge \cdots \ge r^{(S)}$. Furthermore, by replacing some sets $A_i^{(s)}$ with images of themselves under a suitable member of the measure preserving action $\{T_{\mathbf{n}}^{(s)}\}$, we may assume that all of the polynomials under consideration have zero constant term.

Let $X = X^{(1)} \times \cdots \times X^{(s)}$, $\mathcal{B} = \mathcal{B}^{(1)} \otimes \cdots \otimes \mathcal{B}^{(s)}$, and $\mu = \mu^{(1)} \times \cdots \times \mu^{(s)}$. Put $r = r^{(1)}$. For every s with $2 \le s \le S$ for which $r^{(s)} < r$, let $T_{r^{(s)}+1}^{(s)} = T_{r^{(s)}+2}^{(s)} = \cdots = T_r^{(s)}$ be the identity map on $X^{(s)}$. and let

$$p_{r^{(s)}+1,j}^{(s)}(x_1, \cdots, x_k) = p_{r^{(s)}+2,j}^{(s)}(x_1, \cdots, x_k) = \cdots = p_{r,j}^{(s)} = 0, \ 1 \le j \le t^{(s)}.$$

Let P be the set of distinct r-tuples

$$\mathbf{p}_{j,s}(x_1, \cdots x_k) = (p_{1,j}^{(s)}(x_1, \cdots, x_k), p_{2,j}^{(s)}(x_1, \cdots, x_k), \cdots p_{r,j}^{(s)}(x_1, \cdots, x_k))$$

which occur over all values of j and s. For each $\mathbf{p} \in P$ and each s, $1 \le s \le S$, let $A_{\mathbf{p},s}$ be equal to $A_j^{(s)}$ if it happens that $\mathbf{p} = \mathbf{p}_{j,s}$ for some j. Otherwise, let $A_{\mathbf{p},s} = X^{(s)}$. Now let $A_{\mathbf{p}} = A_{\mathbf{p},1} \times \cdots \times A_{\mathbf{p},S}$, $\mathbf{p} \in P$, and put $A = A_0^{(1)} \times \cdots \times A_0^{(s)}$. Also let $T_i = T_i^{(1)} \times \cdots \times T_i^{(s)}$, $1 \le i \le r$, and write, for $\mathbf{n} = (n_1, \cdots, n_r)$, $T_{\mathbf{n}} = T_1^{n_1} \cdots T_r^{n_r}$.

The measure preserving $\mathbf{Z}^r$-action $\{T_{\mathbf{n}}\}_{\mathbf{n} \in \mathbf{Z}^r}$ is not necessarily totally weakly mixing on X, so we are not in a position to utilize directly Theorem D from [BL1], as we did in the proof of Theorem 6.4. However, using the weakly mixing properties of various subgroups of the group $\{T_{\mathbf{n}}\}$ on the product space X and on some of its natural factors, one may still show (see Appendix, Corollary A10) that there exists $(l_1, \cdots, l_k) \in \mathbf{Z}^k$ such that

$$\mu\Big(A_0 \cap \bigcap_{\mathbf{p} \in P} T_{\mathbf{p}(l_1, \cdots, l_k)}^{-1} A_{\mathbf{p}} \Big) > 0.$$

One now proceeds exactly as in the previous proof to conclude that the set

$$R = \Big\{ (n_1, \cdots, n_k) \in \mathbf{Z}^k : \mu\Big(A_0 \cap \bigcap_{\mathbf{p} \in P} T_{\mathbf{p}(n_1, \cdots, n_k)}^{-1} A_{\mathbf{p}} \Big) > 0 \Big\}$$

is a PE-IP*-set. Finally, one may check that $\bigcap_{s=1}^{S} R_{A_0^{(s)},\cdots,A_{t(s)}^{(s)}} \subset R$. Therefore since R is a PE-IP*-set, $\bigcap_{s=1}^{S} R_{A_0^{(s)},\cdots,A_{t(s)}^{(s)}}$ is a PE-IP$^*_+$-set as well.

□

We now turn our attention to mild mixing. This notion is discussed at length in [F2, Chapter 9], and our results here are elaborations on the results there. Recall that a measure preserving $\mathbf{Z}^r$-action $\{T_{\mathbf{n}}\}_{\mathbf{n}\in\mathbf{Z}^r}$ on a probability space $(X,\mathcal{A},\mu)$ is mildly mixing if there are no non-constant rigid functions in $L^2(X,\mathcal{A},\mu)$ ($f \in L^2(X,\mathcal{A},\mu)$ is rigid if there exist a sequence $(\mathbf{n}_k)_{k=1}^{\infty} \subset \mathbf{Z}^r$ with $|\mathbf{n}_k| \to \infty$ such that $T_{\mathbf{n}_k} f \to f$ as $k \to \infty$, where $|\mathbf{v}|$ is the magnitude of a vector $\mathbf{v} \in \mathbf{Z}^r$). The following lemma gives an alternative characterization of mild mixing. For its formulation, we adopt the following notion: an IP-set $(\mathbf{n}_\alpha)_{\alpha\in\mathcal{F}} \subset \mathbf{Z}^r$ will be called *non-trivial* if for no IP-ring $\mathcal{F}^{(1)}$ do we have

$$\underset{\alpha\in\mathcal{F}^{(1)}}{\text{IP-lim}}\ \mathbf{n}_\alpha = 0.$$

Equivalently, $(\mathbf{n}_\alpha)_{\alpha\in\mathcal{F}} \subset \mathbf{Z}^r$ is non-trivial if it is not identically zero along any IP-ring.

For a special case of the following see [F2], (9.11).

Lemma 6.6 A system $(X,\mathcal{A},\mu,\{T_{\mathbf{n}}\}_{\mathbf{n}\in\mathbf{Z}^r})$ is mildly mixing if and only if for every non-trivial IP-set $(\mathbf{n}_\alpha)_{\alpha\in\mathcal{F}} \subset \mathbf{Z}^r$ and every $f, g \in L^2(X,\mathcal{A},\mu)$ there exists an IP-ring $\mathcal{F}^{(1)}$ such that

$$\underset{\alpha\in\mathcal{F}^{(1)}}{\text{IP-lim}} \int f T_{\mathbf{n}_\alpha} g\, d\mu = \Big(\int f\, d\mu\Big)\Big(\int g\, d\mu\Big). \tag{6.3}$$

Proof. Suppose that the system $(X,\mathcal{A},\mu,\{T_{\mathbf{n}}\}_{\mathbf{n}\in\mathbf{Z}^r})$ is mildly mixing, that $(\mathbf{n}_\alpha)_{\alpha\in\mathcal{F}}$ is an IP-set in $\mathbf{Z}^r$, and that $f, g \in L^2(X,\mathcal{A},\mu)$. By confining ourselves to the factor generated by the orbits of f and g under $\{T_{\mathbf{n}}\}$, we may assume that $(X,\mathcal{A},\mu)$ is separable. By Theorem 2.14, therefore, there exists an IP-ring $\mathcal{F}^{(1)}$ such that

$$\underset{\alpha\in\mathcal{F}^{(1)}}{\text{IP-lim}}\ T_{\mathbf{n}_\alpha} g = Pg$$

exists weakly for all $g \in L^2(X,\mathcal{B},\mu)$. It is not hard to show that since $(\mathbf{n}_\alpha)_{\alpha\in\mathcal{F}}$ is non-trivial we may also require of $\mathcal{F}^{(1)}$ that

$$\underset{\alpha\in\mathcal{F}^{(1)}}{\text{IP-lim}}\ |\mathbf{n}_\alpha| = \infty.$$

By Theorem 2.17, P is an orthogonal projection. Let $g \in L^2(X,\mathcal{B},\mu)$. Since

$$Pg = P(Pg) = \underset{\alpha\in\mathcal{F}^{(1)}}{\text{IP-lim}}\ T_{\mathbf{n}_\alpha} Pg,$$

Pg is a rigid function. It follows that Pg is constant for all $g \in L^2(X,\mathcal{B},\mu)$. Therefore, $Pg = \int g\, d\mu$ and (6.3) holds.

For the converse, simply note that if $(X,\mathcal{A},\mu,\{T_{\mathbf{n}}\}_{\mathbf{n}\in\mathbf{Z}^r})$ is not mildly mixing then there exists a non-constant, real-valued function $f \in L^2(X,\mathcal{A},\mu)$ and a sequence $(\mathbf{n}_k)_{k\in\mathbf{N}}$ such that $T_{\mathbf{n}_k} f \to f$ as $k \to \infty$ and such that $|\mathbf{n}_k| > 2\mathbf{n}_{k-1}$, $k = 2, 3, \cdots$. By passing to a subsequence we may assume that $\big|\big|T_{\mathbf{n}_k} f - f\big|\big| < 2^{-k}$ for all $k \in \mathbf{N}$. Letting $\mathbf{n}_\alpha = \sum_{k\in\alpha} \mathbf{n}_k$, $\alpha \in \mathcal{F}$, we have

$$\underset{\alpha\in\mathcal{F}}{\text{IP-lim}}\ T_{\mathbf{n}_\alpha} f = f$$

in the weak topology. Therefore

$$\text{IP-}\lim_{\alpha\in\mathcal{F}} \int fT_{\mathbf{n}_\alpha} f\, d\mu = \Big(\int f^2\, d\mu\Big) > \Big(\int f\, d\mu\Big)^2.$$

Furthermore, by the properties ascribed to the sequence $\mathbf{n}_k$, the IP-set $(\mathbf{n}_\alpha)_{\alpha\in\mathcal{F}}$ is non-trivial. This is a contradiction. Therefore $(X, \mathcal{A}, \mu, \{T_{\mathbf{n}}\}_{\mathbf{n}\in\mathbf{Z}^r})$ is mildly mixing.

□

Our goal is to generalize Lemma 6.6 in a manner similar to Theorem 4.8 of [B1], which states that if T is a mildly mixing measure preserving transformation on $(X, \mathcal{A}, \mu)$, $f_0, f_1, \cdots, f_t \in L^\infty(X, \mathcal{A}, \mu)$, and $p_0(x), p_1(x), \cdots, p_t(x) \in \mathbf{Z}(x)$ are polynomials no two of which differ by a constant then for any $\epsilon > 0$ the set

$$\{0\} \cup \Big\{n \in \mathbf{Z} : \Big|\int \prod_{i=0}^{t} T^{p_i(n)} f_i\, d\mu - \prod_{i=0}^{t} \int f_i\, d\mu\Big| < \epsilon\Big\}$$

is IP*. We must be careful, however, about the IP* property when working in $\mathbf{Z}^k$, $k > 1$. For example, suppose that T is mildly mixing, $p(x_1, x_2) = x_1 - x_2$, and $f \in L^\infty(X, \mathcal{A}, \mu)$ is real valued with $\epsilon = ||f||_2^2 - \big(\int f\, d\mu\big)^2 > 0$. Then the set

$$\Big\{(n_1, n_2) \in \mathbf{Z}^2 : \Big|\int fT^{p(n_1,n_2)} f\, d\mu - \Big(\int f\, d\mu\Big)^2\Big| < \epsilon\Big\}$$

is not an IP*-set in $\mathbf{Z}^2$, for if $(\mathbf{n}_\alpha)_{\alpha\in\mathcal{F}}$ is an IP-set in $\mathbf{Z}^2$ of the form $\mathbf{n}_\alpha = (k_\alpha, k_\alpha)$, where $(k_\alpha)_{\alpha\in\mathcal{F}}$ is an IP-set in $\mathbf{Z}$, we will have $\big|\int fT^{p(\mathbf{n}_\alpha)} f\, d\mu - \big(\int f\, d\mu\big)^2\big| = \epsilon$ for all $\alpha \in \mathcal{F}$.

The obstruction to attaining the right limit in the example above in spite of the presence of mild mixing was the degeneracy of the IP-set $(\mathbf{n}_\alpha)_{\alpha\in\mathcal{F}}$. This provides the impetus for the following definition:

Definition 6.7

a. Suppose that $\mathcal{F}^{(1)}$ is an IP-ring. IP-sets $(n_\alpha^{(1)})_{\alpha\in\mathcal{F}}, \cdots, (n_\alpha^{(k)})_{\alpha\in\mathcal{F}}$ in $\mathbf{Z}$ are said to be *linearly independent along* $\mathcal{F}^{(1)}$ if for every IP ring $\mathcal{F}^{(2)} \subset \mathcal{F}^{(1)}$ and every k integers $l_1, \cdots, l_k$, not all of which are zero, there exists $\alpha \in \mathcal{F}^{(2)}$ such that $l_1 n_\alpha^{(1)} + \cdots + l_k n_\alpha^{(k)} \neq 0$.

b. An IP-set $(\mathbf{n}_\alpha)_{\alpha\in\mathcal{F}} \subset \mathbf{Z}^k$ will be called a *non-degenerate IP-set*, or *NIP-set*, if its k coordinate IP-sets are linearly independent along $\mathcal{F}$. Any subset $E \subset \mathbf{Z}^k$ which intersects non-trivially every NIP-set will be called an *NIP*-set.*

We now have two preparatory lemmas to give before stating our main result concerning polynomial mild mixing of all orders. The first of these generalizes Theorem 9.23 from [F2] and states that mild mixing of $\mathbf{Z}^r$-actions is preserved by the taking of Cartesian products. Its proof, which is not appreciably different from the one in [F2], is supplied for completeness. The second lemma states that linear non-degeneracy of IP-sets implies polynomial non-degeneracy.

Lemma 6.8 The product of two mildly mixing systems $(X, \mathcal{B}, \mu, \{T_{\mathbf{n}}\}_{\mathbf{n}\in\mathbf{Z}^r})$ and $(Y, \mathcal{C}, \nu, \{S_{\mathbf{n}}\}_{\mathbf{n}\in\mathbf{Z}^r})$ is mildly mixing.

Proof. We will use the characterization of Lemma 6.6. Namely, let $(\mathbf{n}_\alpha)_{\alpha\in\mathcal{F}}$ be a non-trivial IP-set in $\mathbf{Z}^r$ and let $f, g \in L^2(X \times Y, \mathcal{B} \otimes \mathcal{C}, \mu \times \nu)$. We must show that there exists an IP-ring $\mathcal{F}^{(1)}$ such that

$$\text{IP-}\lim_{\alpha\in\mathcal{F}^{(1)}} \int f(T_{\mathbf{n}_\alpha} \times S_{\mathbf{n}_\alpha}) g\, d\mu \times \nu = \Big(\int f\, d\mu \times \nu\Big)\Big(\int g\, d\mu \times \nu\Big). \tag{6.4}$$

Write $f = \sum \phi_i \otimes \psi_i$ and $g = \sum \phi_j' \otimes \psi_j'$, where the sum is countable. (Here as always, $\phi \otimes \psi(x,y) = \phi(x)\psi(y)$.)

By Proposition 2.14 and the fact that both systems are mildly mixing, we may choose an IP-ring $\mathcal{F}^{(1)}$ such that

$$\underset{\alpha\in\mathcal{F}^{(1)}}{\text{IP-lim}} \int \phi_i T_{\mathbf{n}_\alpha} \phi_j' \, d\mu = \Big(\int \phi_i \, d\mu\Big)\Big(\int \phi_j' \, d\mu\Big)$$

and

$$\underset{\alpha\in\mathcal{F}^{(1)}}{\text{IP-lim}} \int \psi_i S_{\mathbf{n}_\alpha} \psi_j' \, d\nu = \Big(\int \psi_i \, d\nu\Big)\Big(\int \psi_j' \, d\nu\Big)$$

for all i, j.

We now have

$$\begin{aligned}
&\underset{\alpha\in\mathcal{F}^{(1)}}{\text{IP-lim}} \int f(T_{\mathbf{n}_\alpha} \times S_{\mathbf{n}_\alpha}) g \, d\mu \times \nu \\
=&\underset{\alpha\in\mathcal{F}^{(1)}}{\text{IP-lim}} \sum_{i,j} \int\int \phi_i(x) T_{\mathbf{n}_\alpha}\phi_j'(x)\psi_i(y) S_{\mathbf{n}_\alpha}\psi_j'(y) \, d\mu(x) d\nu(y) \\
=&\underset{\alpha\in\mathcal{F}^{(1)}}{\text{IP-lim}} \sum_{i,j} \Big(\int \phi_i T_{\mathbf{n}_\alpha} \phi_j' \, d\mu\Big)\Big(\int \psi_i S_{\mathbf{n}_\alpha} \psi_j' \, d\nu\Big) \\
=&\sum_{i,j} \Big(\int \phi \, d\mu\Big)\Big(\int \phi' \, d\mu\Big)\Big(\int \psi \, d\nu\Big)\Big(\int \psi' \, d\nu\Big) \\
=&\Big(\sum_i \Big(\int \phi_i \, d\mu\Big)\Big(\int \psi_i \, d\mu\Big)\Big)\Big(\sum_j \Big(\int \phi_j' \, d\mu\Big)\Big(\int \psi_j' \, d\mu\Big)\Big) \\
=&\Big(\int f \, d\mu \times \nu\Big)\Big(\int g \, d\mu \times \nu\Big).
\end{aligned}$$

□

Here is the lemma about polynomial degeneracy.

Lemma 6.9 Suppose that $\mathcal{F}^{(1)}$ is an IP-ring and $(n_\alpha^{(1)})_{\alpha\in\mathcal{F}}, \cdots, (n_\alpha^{(k)})_{\alpha\in\mathcal{F}}$ are IP-sets in $\mathbf{Z}$ which are linearly independent along $\mathcal{F}^{(1)}$. Then

a. For any polynomial $p(x_1, \cdots, x_k) \in \mathbf{Z}[x_1, \cdots, x_k]$ we have

$$\underset{\alpha\in\mathcal{F}^{(1)}}{\text{IP-lim}} \ p(n_\alpha^{(1)}, \cdots, n_\alpha^{(k)}) = 0 \tag{6.5}$$

only if $p(x_1, \cdots, x_k) = 0$.

b. If $p(x_1, \cdots, x_k) \in \mathbf{Z}[x_1, \cdots, x_k]$ is not constant then for some IP-ring $\mathcal{F}^{(2)} \subset \mathcal{F}^{(1)}$ we have

$$\underset{\alpha\in\mathcal{F}^{(2)}}{\text{IP-lim}} \ |p(n_\alpha^{(1)}, \cdots, n_\alpha^{(k)})| = \infty. \tag{6.6}$$

Proof. The proof of part a. is by induction on the degree of p. The conclusion is true by definition if the degree of p is at most 1. Suppose now that the conclusion is valid for all p of degree at most $d-1$ and assume that p is a polynomial of degree at most d satisfying (6.5).

There exists $\alpha_0 \in \mathcal{F}$ such that for all $\alpha \in \mathcal{F}^{(1)}$ with $\alpha > \alpha_0$, $p(n_\alpha^{(1)}, \cdots, n_\alpha^{(k)}) = 0$. Fix $\alpha \in \mathcal{F}^{(1)}$ with $\alpha > \alpha_0$ and let

$$q(x_1, \cdots, x_k) = p(x_1 + n_\alpha^{(1)}, \cdots, x_k + n_\alpha^{(k)}) - p(x_1, \cdots, x_k) - p(n_\alpha^{(1)}, \cdots, n_\alpha^{(k)}). \tag{6.7}$$

Then $\deg q = \max\{\deg p - 1, 0\}$.

For all $\beta \in \mathcal{F}^{(1)}$ with $\beta > \alpha$ we have $\alpha \cup \beta \in \mathcal{F}^{(1)}$ and $\alpha \cup \beta > \alpha_0$, so

$$p(n_\alpha^{(1)} + n_\beta^{(1)}, \cdots, n_\alpha^{(k)} + n_\beta^{(k)}) = p(n_{\alpha\cup\beta}^{(1)}, \cdots, n_{\alpha\cup\beta}^{(k)}) = 0.$$

Using (6.7), coupled with the fact that $p(n_\alpha^{(1)}, \cdots, n_\alpha^{(k)}) = p(n_\beta^{(1)}, \cdots, n_\beta^{(k)}) = 0$, we may conclude that $q(n_\beta^{(1)}, \cdots, n_\beta^{(k)}) = 0$ for all $\beta \in \mathcal{F}^{(1)}$ with $\beta > \alpha$. By hypothesis $q(x_1, \cdots, x_k) = 0$ and $\deg p \leq 1$, which implies that $p(x_1, \cdots, x_k) = 0$ by an earlier case.

For part b., note that by Proposition 2.14 we may choose an IP-ring $\mathcal{F}^{(2)} \subset \mathcal{F}^{(1)}$ such that either the limit

$$\text{IP-}\lim_{\alpha \in \mathcal{F}^{(2)}} |p(n_\alpha^{(1)}, \cdots, n_\alpha^{(k)})| = M \tag{6.8}$$

exists and is finite or so that (6.6) holds. We need merely show that the former case is impossible. Namely, assume that we have (6.8). Letting $q(x_1, \cdots, x_k) = p(x_1, \cdots, x_k) - M$, we have

$$\text{IP-}\lim_{\alpha \in \mathcal{F}^{(2)}} q(n_\alpha^{(1)}, \cdots, n_\alpha^{(k)}) = 0,$$

so that by part a. $q(x_1, \cdots, x_k) = 0$. It follows that $p(x_1, \cdots, x_k)$ is constant, a contradiction.

□

Armed with these two lemmas, we are now able to demonstrate that the following theorem is just an absolute case of Theorem 4.10.

Theorem 6.10 Suppose that $(X, \mathcal{A}, \mu)$ is a probability space and $\{T_{\mathbf{n}}\}_{\mathbf{n}\in\mathbf{Z}^r}$ is a mild mixing measure preserving $\mathbf{Z}^r$-action on X generated by $T_1, \cdots, T_r$. Suppose that $t \in \mathbf{N}$ and $p_{i,j}(x_1, \cdots, x_k) \in \mathbf{Q}[x_1, \cdots, x_k]$ with $p_{i,j}(\mathbf{Z}^k) \subset \mathbf{Z}$, $1 \leq i \leq r$, $1 \leq j \leq t$ such that for any $1 \leq j_1 \neq j_2 \leq t$, the functions

$$(l_1, \cdots, l_k) \to \big(p_{1,j_1}(l_1, \cdots, l_k), \cdots, p_{r,j_1}(l_1, \cdots, l_k)\big)$$

and

$$\begin{aligned}&(l_1, \cdots, l_k)\\ \to &\big(p_{1,j_1}(l_1, \cdots, l_k) - p_{1,j_2}(l_1, \cdots, l_k), \cdots, p_{r,j_1}(l_1, \cdots, l_k) - p_{r,j_2}(l_1, \cdots, l_k)\big)\end{aligned}$$

are not constant. Suppose $f_0, f_1, \cdots, f_t \in L^\infty(X, \mathcal{A}, \mu)$ and let $p_{i,0}(x_1, \cdots, x_k) = 0$, $1 \leq i \leq r$. Then for any $\epsilon > 0$ the set

$$S_\epsilon = \left\{(n_1, \cdots, n_k) \in \mathbf{Z}^k : \prod_{j=0}^t \Big(\prod_{i=1}^r T_i^{p_{i,j}(n_1, \cdots, n_k)}\Big) f_j - \prod_{j=0}^t \int f_j \, d\mu| < \epsilon\right\}$$

is an NIP*-set.

Proof. The first observation is that we may assume as always and without loss of generality that $(X, \mathcal{A}, \mu)$ is separable and that $p_{i,j}(x_1, x_2, \cdots, x_k) \in \mathbf{Z}[x_1, \cdots, x_k]$, $1 \leq i \leq r$, $1 \leq j \leq t$. Next, we note that we may assume that $p_{i,j}(0, 0, \cdots, 0) = 0$.

(All that is involved here is replacing the set A_j by some set from its orbit under the action, namely $T_{\mathbf{n}}^{-1}A_j$, where $\mathbf{n} = \big(p_{1,j}(0,\cdots 0),\cdots,p_{r,j}(0,\cdots,0)\big)$.)

According to Lemma 6.7, the product of the action $\{T_{\mathbf{n}}\}_{\mathbf{n}\in\mathbf{Z}^r}$ with itself on the product space $(X\times X,\mathcal{A}\otimes\mathcal{A},\mu\times\mu)$ is mildly mixing. We will denote this product action by the same symbols $\{T_{\mathbf{n}}\}$.

Let $\epsilon > 0$ and let $(\mathbf{n}_\alpha)_{\alpha\in\mathcal{F}}$ be any NIP-set in $\mathbf{Z}^k$ This IP-set has the form $\mathbf{n}_\alpha = (n_\alpha^{(1)},\cdots,n_\alpha^{(k)})$, where $(n_\alpha^{(i)})_{\alpha\in\mathcal{F}}$, $1\leq i\leq k$ are linearly independent IP-sets in $\mathbf{Z}$. Let

$$d = \max_{1\leq i\leq r,\ 1\leq j\leq t} p_{i,j}$$

and let $\mathbf{PE}_{\leq d}(1)$ be as in Definition 1.1.

By Proposition 2.17 we may choose an IP-ring $\mathcal{F}^{(1)}$ such that

$$\operatorname*{IP-lim}_{\alpha\in\mathcal{F}^{(1)}} S(\alpha)H = P_S H$$

for all $H\in L^2(X\times X,\mathcal{A}\otimes\mathcal{A},\mu\times\mu)$ and all $S\in\mathbf{PE}_{\leq d}(1)$. According to Theorem 2.17, P_S is an orthogonal projection for each S. Suppose $H\in L^2(X\times X,\mathcal{A}\otimes\mathcal{A},\mu\times\mu)$ and $S\in\mathbf{PE}_{\leq d}$ is not the identity element. Define an $\mathcal{F}$-sequence $(\mathbf{t}_\alpha)_{\alpha\in\mathcal{F}}$ by $S(\alpha) = T^{\mathbf{t}_\alpha}$. We have

$$\operatorname*{IP-lim}_{\alpha\in\mathcal{F}^{(1)}} T_{\mathbf{t}_\alpha}\big(P_S H\big) = \operatorname*{IP-lim}_{\alpha\in\mathcal{F}^{(1)}} S(\alpha)\big(P_S H\big) = P_S^2 H = P_S H. \tag{6.9}$$

It is a consequence of Lemma 6.9 b. that for some IP-ring $\mathcal{F}^{(2)}\subset\mathcal{F}^{(1)}$ we have

$$\operatorname*{IP-lim}_{\alpha\in\mathcal{F}^{(2)}} |\mathbf{t}_\alpha| = \infty.$$

This fact together with (6.9) implies that $P_S H$ is a rigid function and therefore constant. Since H was arbitrary, this in turn implies that S is mixing on X over the trivial factor Y along $\mathcal{F}^{(1)}$ for all $S\in\mathbf{PE}_{\leq d}(1)$ different from the identity. Therefore, if we let $G = \{I\}$, G is a subgroup of $\mathbf{PE}_{\leq d}(1)$ and it is not difficult to show that X is a primitive extension of Y along $\mathcal{F}^{(1)}$ with compact part G. Let

$$S_j = \prod_{i=1}^{r} T_i^{p_{i,j}(n_\alpha^{(1)},\cdots,n_\alpha^{(k)})},\ 1\leq j\leq t.$$

The conditions of the theorem ensure that $S_{j_1}\neq I$ and $S_{j_1}S_{j_2}^{-1}\neq I$ when $1\leq j_1\neq j_2\leq t$. Therefore the set $\{S_1,\cdots,S_t\}$ is a mixing set in accordance with Definition 3.5.

Set $S_0 = I$. Remembering that Y is the trivial factor, we have by Theorem 4.10 (1) that there exists an IP-ring $\mathcal{F}^{(2)}\subset\mathcal{F}^{(1)}$ having the property that

$$\operatorname*{IP-lim}_{\alpha\in\mathcal{F}^{(2)}} \left(\int\prod_{i=0}^{t} S_i(\alpha)f_i\,d\mu = \prod_{i=0}^{t}\int f_i\,d\mu\right) = 0.$$

Clearly this yields $\alpha\in\mathcal{F}^{(1)}$ such that $\mathbf{n}_\alpha\in S_\epsilon$.

□

The following fact, which was mentioned in the introduction, is needed for our next application. It's easy proof is included for the sake of completeness.

Proposition 6.11 Any IP*-set in $\mathbf{Z}^r$ is syndetic.

Proof. Suppose $E \subset \mathbf{Z}^r$ is not syndetic. We will show that E is not IP* by exhibiting an IP-set contained in E^c. Let $\mathbf{n}_1 \in E^c$. Since E is not syndetic, $\big(E \cup (E - \mathbf{n}_1))\big) \neq \mathbf{Z}^r$, that is, there exists $\mathbf{n}_2 \in \big(E^c \cap (E - \mathbf{n}_1)^c\big)$. Then $\{\mathbf{n}_1, \mathbf{n}_2, \mathbf{n}_1 + \mathbf{n}_2\} \subset E^c$. Again since E is not syndetic,

$$\big(E \cup (E - \mathbf{n}_1) \cup (E - \mathbf{n}_2) \cup (E - (\mathbf{n}_1 + \mathbf{n}_2))\big) \neq \mathbf{Z}^r,$$

so we may find

$$\mathbf{n}_3 \in \big(E^c \cap (E - \mathbf{n}_1)^c \cap (E - \mathbf{n}_2)^c \cap (E - (\mathbf{n}_1 + \mathbf{n}_2))^c\big).$$

Then

$$\{\mathbf{n}_1, \mathbf{n}_2, \mathbf{n}_1 + \mathbf{n}_2, \mathbf{n}_3, \mathbf{n}_3 + \mathbf{n}_1, \mathbf{n}_3 + \mathbf{n}_2, \mathbf{n}_3 + \mathbf{n}_2 + \mathbf{n}_1\} \subset E^c.$$

Continuing in this fashion we get a sequence $(\mathbf{n}_i)_{i=1}^\infty$ such that the IP-set $FS(\mathbf{n}_i)_{i=1}^\infty$ is contained in E^c.

□

At this time we would like to tie up one loose end, which is to show that a uniform version of Theorem 0.4 may be obtained as a corollary to Theorem 1.3 (By "uniform" we mean replacing the limit $\lim_{N\to\infty} \frac{1}{N} \cdots$ by a limit of the type $\lim_{N-M\to\infty} \frac{1}{N-M} \cdots$.) It is not immediately obvious how one would obtain such a result from, for example, Theorem 0.9. First, we have a lemma.

Lemma 6.12 Suppose that we have r commuting measure preserving transformations $T_1, \cdots, T_r$ of a probability space $(X, \mathcal{B}, \mu)$. Suppose $k, t \in \mathbf{N}$, and $p_{i,j}(n_1, \cdots, n_k) \in \mathbf{Q}[n_1, \cdots, n_k]$ with $p_{i,j}(\mathbf{Z}^k) \subset \mathbf{Z}$ and $p_{i,j}(0, \cdots, 0) = 0$, $1 \le i \le r$, $1 \le j \le t$. Then for every $A \in \mathcal{B}$ with $\mu(A) > 0$ there exists $\epsilon > 0$ such that the set

$$R_{A,\epsilon} = \left\{(n_1, \cdots, n_k) \in \mathbf{Z}^k : \mu\bigg(\bigcap_{j=1}^{t} \Big(\prod_{i=1}^{r} T_i^{p_{i,j}(n_1,\cdots,n_k)}\Big)^{-1} A\bigg) > \epsilon\right\} \tag{6.10}$$

is a syndetic set in $\mathbf{Z}^k$.

Proof. Suppose to the contrary that for all $n \in \mathbf{N}$ the set $R_{A,\frac{1}{n}}$ fails to be syndetic. Then for every $n \in \mathbf{N}$ we may find a k-dimensional parallelepiped $I_n = [M_{n,1}, N_{n,1}] \times \cdots \times [M_{n,k}, N_{n,k}]$ such that

$$\lim_{n\to\infty} (N_{n,i} - M_{n,i}) = 0, \ 1 \le i \le k$$

and such that for all $(n_1, \cdots, n_k) \in I_n$ we have

$$\mu\bigg(\bigcap_{j=1}^{t} \Big(\prod_{i=1}^{r} T_i^{p_{i,j}(n_1,\cdots,n_k)}\Big)^{-1} A\bigg) < \frac{1}{n}.$$

The complement of the set $\bigcup_{n=1}^\infty I_n$ is clearly not syndetic. By Lemma 6.11, it is therefore not IP*. Hence there exists an IP-set $\big((n_\alpha^{(1)}, \cdots, n_\alpha^{(k)})\big)_{\alpha\in\mathcal{F}}$ contained in $\bigcup_{n=1}^\infty I_n$. By passing to a subring $\mathcal{F}^{(1)}$, we may assume that

$$\operatorname*{IP-lim}_{\alpha\in\mathcal{F}^{(1)}} |(n_\alpha^{(1)}, \cdots, n_\alpha^{(k)})|$$

exists as either a finite or infinite limit. Since the only finite limit possible is clearly 0, this limit must be infinite, which implies by construction that

$$\operatorname*{IP-lim}_{\alpha\in\mathcal{F}^{(1)}} \mu\Big(\bigcap_{j=1}^{t}\Big(\prod_{i=1}^{r} T_i^{p_{i,j}(n_\alpha^{(1)},\cdots,n_\alpha^{(k)})}\Big)^{-1}A\Big)=0.$$

This contradicts Theorem 1.3, which indicates that the limit must be positive.

□

The single parameter case $k=1$ of the following multi-parameter theorem gives the uniform version of Theorem 0.4 promised in the introduction.

Theorem 6.13 Suppose we are given r commuting invertible measure preserving transformations $T_1,\cdots,T_r$ of a probability space $(X,\mathcal{B},\mu)$. Let $k,t\in\mathbf{N}$, and suppose that $p_{i,j}(n_1,\cdots,n_k)\in\mathbf{Q}[n_1,\cdots,n_k]$ with $p_{i,j}(\mathbf{Z}^k)\subset\mathbf{Z}$ and $p_{i,j}(0,\cdots,0)=0$, $1\le i\le r$, $1\le j\le t$. Then for every $A\in\mathcal{B}$ with $\mu(A)>0$ we have

$$\liminf_{\substack{N_i-M_i\to\infty\\ 1\le i\le r}}\frac{1}{\prod_{i=1}^{r}(N_i-M_i)}\sum_{\substack{n_i\in[M_i+1,N_i]\\ 1\le i\le r}}\mu\Big(\bigcap_{j=1}^{t}\Big(\prod_{i=1}^{r}T_j^{p_{i,j}(n_1,\cdots,n_k)}\Big)A\Big)>0.$$

Proof. By Lemma 6.12, there exists ϵ such that the set $R_{A,\epsilon}$ defined by (6.10) is syndetic. This implies that for some k-dimensional cube I we have $I+R_{A,\epsilon}=\mathbf{Z}^k$. Since any big enough k-dimensional parallelepiped may be approximately tiled by shifts of I, the limit appearing in (6.11) must have value no less than $\frac{\epsilon}{|I|}$.

□

Lemma 6.12 suggests the following question, which in the non-linear case we do not know the answer to.

Question 2. Under the conditions of Lemma 6.12, does there exist $\epsilon>0$ such that the set $R_{A,\epsilon}$ defined by (6.10) is IP*?

In the linear case the answer is yes. This may be shown using Furstenberg's and Katznelson's IP-multiple recurrence theorem for countably generated IP-systems, as we shall now demonstrate via a version of a combinatorial corollary to their theorem which we shall state presently. For its formulation we introduce the following notation: for $t,s,l\in\mathbf{N}$ let $M(t,s,l)$ denote the set of all $t\times s$ matrices $(m_{i,j})$ whose entries are taken from $\{1,2,\cdots,l\}$. Note that $|M(t,s,l)|=l^{ts}$. For $\alpha\in FU\{1,2,\cdots,l\}$ and $1\le i\le t$, let $e_\alpha^{(i)}$ be the $t\times s$ matrix $(m_{a,b})$ defined by $m_{a,b}=1$ if $a=i$ and $b\in\alpha$ and $m_{a,b}=0$ otherwise. Here now is Theorem 9.2 from [FK1].

Theorem 6.14 Let $t\in\mathbf{N}$ and $\delta>0$ be given. There exist $l_0=l_0(t,\delta)$ and $s_0=s_0(t,\delta)$ in $\mathbf{N}$ having the property that for every $l\ge l_0$, every $s\ge s_0$, and every subset $S\subset M(t,s,l)$ satisfying $|S|\ge\delta l^{ts}$ there exist $\alpha\in FU\{1,\cdots,s\}$ and a matrix $A\in M(t,s,l)$ such that

$$\{A,A+e_\alpha^{(1)},\cdots,A+e_\alpha^{(t)}\}\subset S.$$

With this in hand we can easily get a positive answer to Question 2 in the linear case, and in fact more. Following [FK1], let us say that a set in $\mathbf{Z}^k$ which is of the form $\{n_\alpha : \alpha \in FU\{1,2,\cdots,s\}\}$, where $n_{\alpha\cup\beta} = n_\alpha + n_\beta$ when $\alpha\cap\beta = \emptyset$, is an *$IP_s$-set.* (Thus, an IP_s-set is just a finite IP-set with s generators.) Accordingly, we shall call a subset E of $\mathbf{Z}^k$ an *IP^*_s-set* if it intersects non-trivially every IP_s-set. Since every IP-set contains (infinitely many) IP_s-sets, it is clear that every IP^*_s-set is also an IP^*-set. However, for every s there exist IP^*-sets which are not IP^*_s-sets. For example, by [FK1], Theorem 10.4 the lower density of any IP^*_s-set in $\mathbf{Z}$ is at least 2^{-s+1}, whereas $k\mathbf{Z}$ is IP^* for any positive integer k. Therefore, the following Theorem, which deals with linear polynomials, answers Question 2 in the affirmative for the linear case and gives somewhat more as well.

Theorem 6.15 Suppose $t \in \mathbf{N}$ and $\delta > 0$ are given. There exist $s = s(t,\delta) \in \mathbf{N}$ and $\xi = \xi(t,\delta) > 0$ having the property that for all $r, k \in \mathbf{N}$, if r commuting measure preserving transformations $T_1,\cdots,T_r$ of a probability space $(X,\mathcal{B},\mu)$ are given, as well as linear polynomials $p_{i,j}(n_1,\cdots,n_k) \in \mathbf{Q}[n_1,\cdots,n_k]$ with $p_{i,j}(\mathbf{Z}^k) \subset \mathbf{Z}$ and $p_{i,j}(0,\cdots,0) = 0$, $1 \le i \le r$, $1 \le j \le t$, then for every $B \in \mathcal{B}$ with $\mu(B) > \delta$ the set

$$R_{B,\xi} = \left\{(n_1,\cdots,n_k) \in \mathbf{Z}^k : \mu\left(\bigcap_{j=1}^{t}\left(\prod_{i=1}^{r} T_i^{p_{i,j}(n_1,\cdots,n_k)}\right)^{-1} B\right) > \xi\right\} \tag{6.11}$$

is an IP^*_s-set in $\mathbf{Z}^k$.

Proof. Let $s = s_0(t,\frac{\delta}{2})$ and $l = l_0(t,\frac{\delta}{2})$ be as in Theorem 6.14 and let $\xi = \frac{\delta}{2^{s+1}l^{ts}}$. Suppose that k, r, $(X,\mathcal{A},\mu)$, transformations T_i, polynomials $p_{i,j}$, etc. have been given. Suppose now that $B \in \mathcal{B}$ with $\mu(B) > 0$. We claim that the set $R_{B,\xi}$ defined by (6.11) is an IP^*_s-set. Accordingly, let $(\mathbf{n}_\alpha)$ be any IP_s-set in $\mathbf{Z}^s$. Write

$$\mathbf{p}_j(x_1,\cdots,x_k) = \big(p_{1,j}(x_1,\cdots,x_k),\cdots,p_{r,j}(x_1,\cdots,x_k)\big),\ 1 \le j \le t.$$

Since all of the polynomials are linear, $\big(\mathbf{p}_j(\mathbf{n}_\alpha)\big)$ is an IP_s-set in $\mathbf{Z}^r$, $1 \le j \le t$. For $\mathbf{n} = (n_1,\cdots,n_r) \in \mathbf{Z}^r$, write $T_{\mathbf{n}} = T_1^{n_1}\cdots T_r^{n_r}$.

For every matrix $M = (m_{i,j}) \in M(t,s,l)$, let $l_M \in \mathbf{Z}^r$ be defined by

$$l_M = \sum_{1\le j\le t,\ 1\le i\le s} m_{i,j}\mathbf{p}_j(\mathbf{n}_{\{i\}})$$

and put $B_M = T_{l_M}B$. Defined a function f on X by

$$f = \frac{1}{l^{ts}} \sum_{M\in M(t,s,l)} 1_{B_M}$$

and let $C = \{x : f(x) \ge \frac{\delta}{2}\}$. Since $f(X) \subset [0,1]$ and $\int f\, d\mu = \mu(B) \ge \delta$ we have $\mu(C) \ge \frac{\delta}{2}$. We claim that

$$C \subset \bigcup_{A\in M(t,s,l),\ \alpha\in FU(\{1,\cdots,s\})} \big(B_A \cap B_{A+e_\alpha^{(1)}} \cap\cdots\cap B_{A+e_\alpha^{(t)}}\big). \tag{6.12}$$

To see this, consider that for any $x \in C$ the set $S = \{M \in M(t,s,l) : x \in A_M\}$ has cardinality at least $\frac{\delta}{2}l^{ts}$ and therefore by the conclusion of Theorem 6.14 contains a configuration of the form $\{A, A+e_\alpha^{(1)},\cdots,A+e_\alpha^{(t)}\}$.

Since the number of sets in the union appearing in (6.12) is no more than $2^s l^{ts}$, one of the sets in the union, say $B' = B_A \cap B_{A+e_\alpha^{(1)}} \cap \cdots \cap B_{A+e_\alpha^{(t)}}$, must have measure at least $\frac{\delta}{2^{s+1}l^{ts}} = \xi$. One need only now check that

$$T_{l_A}^{-1} B' = \big(B \cap T_{\mathbf{p}_1(\mathbf{n}_\alpha)} \cap \cdots \cap T_{\mathbf{p}_t(\mathbf{n}_\alpha)}\big),$$

which implies that

$$\mu\bigg(\bigcap_{j=1}^{t}\Big(\prod_{i=1}^{r} T_i^{p_{i,j}(\mathbf{n}_\alpha)}\Big)^{-1} B\bigg) > \zeta$$

and we are done.

□

CHAPTER 7

COMBINATORIAL APPLICATIONS

In this section we will give some very straightforward combinatorial applications of the multiple recurrence results obtained in Section 6. These will all be achieved via Furstenberg's correspondence principle. First we remind the reader of the notions of upper and lower Banach density in $\mathbf{Z}^r$.

Definition 7.1 Suppose that $r \in \mathbf{N}$ and $E \subset \mathbf{Z}^r$ is a set. The upper Banach density of E is defined to be the number

$$d^*(E) = \limsup_{N_i - M_i \to \infty,\ 1 \le i \le r} \frac{|E \cap \prod_{i=1}^r \{M_i, M_i + 1, \cdots, N_i - 1\}|}{\prod_{i=1}^r (N_i - M_i)}.$$

The *lower Banach density* of E is defined to be the number

$$d_*(E) = \liminf_{N_i - M_i \to \infty,\ 1 \le i \le r} \frac{|E \cap \prod_{i=1}^r \{M_i, M_i + 1, \cdots, N_i - 1\}|}{\prod_{i=1}^r (N_i - M_i)}.$$

Furstenberg's correspondence principle has many different versions. This one is the one most suitable for our present context. We give a proof for completeness.

Proposition 7.2 Suppose that $r \in \mathbf{N}$ and $E \subset \mathbf{Z}^r$. There exists a measure preserving system $(X, \mathcal{A}, \mu, \{T_{\mathbf{n}}\}_{\mathbf{n} \in \mathbf{Z}^r})$ and some $A \in \mathcal{A}$ with $\mu(A) = d^*(E)$ such that for all $k \in \mathbf{N}$ and all $\mathbf{n}_1, \cdots, \mathbf{n}_k \in \mathbf{Z}^r$ we have

$$d^*\big(E \cap (E - \mathbf{n}_1) \cap \cdots \cap (E - \mathbf{n}_k)\big) \ge \mu\big(A \cap T_{\mathbf{n}_1} A \cap \cdots \cap T_{\mathbf{n}_k} A\big).$$

Proof. Let $X = \{0,1\}^{\mathbf{Z}^r}$ be the set of all functions $\varphi : \mathbf{Z}^r \to \{0,1\}$. Let $T_i : X \to X$ be the right ith coordinate shift:

$$T_i \varphi(z_1, z_2, \cdots, z_r) = \varphi(z_1, \cdots, z_{i-1}, z_i - 1, z_{i+1}, \cdots, z_r).$$

Clearly $T_1, T_2, \cdots, T_r$ are commuting. For $\mathbf{n} = (n_1, \cdots, n_r) \in \mathbf{Z}^r$ put

$$T_{\mathbf{n}} = T_1^{n_1} \cdots T_r^{n_r}.$$

Choose a sequence of parallelepipeds $P_t = \prod_{i=1}^r \{M_i^{(t)}, \cdots, N_i^{(t)} - 1\}$ such that

$$\lim_{t \to \infty} \frac{|E \cap P_t|}{|P_t|} = d^*(E).$$

Let $\xi = 1_E \in X$ and let $A = \{\chi \in X : \chi(0,0,\cdots,0) = 1\}$. Then

$$\lim_{t\to\infty} \frac{1}{|P_t|} \sum_{\mathbf{n}\in P_t} 1_A(T_{\mathbf{n}}\xi) = \lim_{t\to\infty} \frac{1}{|P_t|} \sum_{\mathbf{n}\in P_t} 1_E(\mathbf{n}) = d^*(E).$$

Noting that $\mathcal{C}(X)$ is separable and using a diagonal argument, we may by passing to a subsequence of $(P_t)_{t=1}^{\infty}$, say $(P_{t_s})_{s=1}^{\infty}$, assume that

$$\lim_{s\to\infty} \frac{1}{|P_{t_s}|} \sum_{\mathbf{n}\in P_{t_s}} f(T_{\mathbf{n}}\xi) = L(f)$$

exists for all $f \in \mathcal{C}(X)$. $L(f)$ is a positive linear functional and so by the Riesz representation theorem is given by integration against a Borel probability measure μ. Let $\mathcal{A}$ be the Borel σ-algebra. Note finally that $L(f) = L(T_{\mathbf{n}}f)$ for all $\mathbf{n} \in \mathbf{Z}^r$, so that μ is $T_{\mathbf{n}}$-invariant, $\mathbf{n} \in \mathbf{Z}^r$, and

$$\mu(A) = \int 1_A \, d\mu = L(1_A) = \lim_{t\to\infty} \frac{1}{|P_t|} \sum_{\mathbf{n}\in P_t} 1_A(T_{\mathbf{n}}\xi) = d^*(E).$$

Now we have

$$\begin{aligned}
&\mu(A \cap T_{\mathbf{n}_1} A \cap \cdots \cap T_{\mathbf{n}_k} A) \\
=& \int 1_{A\cap T_{\mathbf{n}_1}A\cap\cdots\cap T_{\mathbf{n}_k}A} \, d\mu \\
=& L(1_{A\cap T_{\mathbf{n}_1}A\cap\cdots\cap T_{\mathbf{n}_k}A}) \\
=& \lim_{s\to\infty} \frac{1}{|P_{t_s}|} \sum_{\mathbf{n}\in P_{t_s}} 1_{A\cap T_{\mathbf{n}_1}A\cap\cdots\cap T_{\mathbf{n}_k}A}(T_{\mathbf{n}}\xi) \\
=& \lim_{s\to\infty} \frac{1}{|P_{t_s}|} \sum_{\mathbf{n}\in P_{t_s}} 1_{E\cap(E-\mathbf{n}_1)\cap\cdots\cap(E-\mathbf{n}_k)}(\mathbf{n}) \\
\le& d^*\big(E \cap (E-\mathbf{n}_1) \cap \cdots \cap (E-\mathbf{n}_k)\big).
\end{aligned}$$

□

With Furstenberg's correspondence principle in hand, we may proceed to give our primary combinatorial application, which is a corollary of Theorem 6.3.

Theorem 7.3 Suppose that $r, k, t \in \mathbf{N}$, $E \subset \mathbf{Z}^r$ with $d^*(E) > 0$, and $p_i : \mathbf{Z}^k \to \mathbf{Z}^r$ are polynomial mappings with $p_i(\mathbf{0}) = \mathbf{0}$, $1 \le i \le t$. Then

$$\Big\{\mathbf{n} \in \mathbf{Z}^k : d^*\Big(E \cap \big(E - p_1(\mathbf{n})\big) \cap \cdots \cap \big(E - p_t(\mathbf{n})\big)\Big) > 0\Big\}$$

is a PE-IP*-set in $\mathbf{Z}^k$.

Proof. By Furstenberg's correspondence principle there exists a measure preserving system $(X, \mathcal{A}, \mu, \{T_{\mathbf{n}}\}_{\mathbf{n}\in\mathbf{Z}^r})$ and $A \in \mathcal{A}$ with $\mu(A) > 0$ such that for all $\mathbf{n}_1, \cdots, \mathbf{n}_k \in \mathbf{Z}^r$ we have

$$d^*\big(E \cap (E-\mathbf{n}_1) \cap \cdots \cap (E-\mathbf{n}_t)\big) \ge \mu(A \cap T_{\mathbf{n}_1} A \cap \cdots \cap T_{\mathbf{n}_t} A).$$

Hence we have

$$\begin{aligned}
&\Big\{\mathbf{n} \in \mathbf{Z}^k : \mu(A \cap T_{p_1(\mathbf{n})} A \cap \cdots \cap T_{p_t(\mathbf{n})} A) > 0\Big\} \\
\subset& \Big\{\mathbf{n} \in \mathbf{Z}^k : d^*\Big(E \cap \big(E - p_1(\mathbf{n})\big) \cap \cdots \cap \big(E - p_t(\mathbf{n})\big)\Big) > 0\Big\}.
\end{aligned}$$

The latter is a PE-IP*-set by Theorem 6.3.

□

Here now is the aforementioned result, which is a more general version of Theorem 0.11 from the introduction.

Theorem 7.4 Let $t, k, r \in \mathbf{N}$ and let $\mathbf{p}_i : \mathbf{Z}^k \to \mathbf{Z}^r$ be polynomial mappings satisfying $\mathbf{p}_i(0, \cdots, 0) = (0, \cdots, 0)$, $1 \leq i \leq t$. Suppose that $s \in \mathbf{N}$ and that $\mathbf{Z}^r = \bigcup_{i=1}^s C_i$ is a partition of $\mathbf{Z}^r$ into s cells. Then there exists some $L \in \mathbf{N}$ and some $\epsilon > 0$ having the property that in any rectangle

$$B = [M_1, N_1] \times [M_2, N_2] \times \cdots \times [M_r, N_r] \subset \mathbf{Z}^r$$

with $\min_{1 \leq i \leq r}(N_i - M_i) \geq L$ there exists i with $1 \leq i \leq s$ and $\mathbf{n} \in C_i \cap B$ such that

$$d^*\Big(C_i \cap (C_i - \mathbf{p}_1(\mathbf{n})) \cap \cdots \cap (C_i - \mathbf{p}_t(\mathbf{n}))\Big) > \epsilon.$$

In particular, the system of polynomial equations

$$\begin{aligned} \mathbf{x}_0 &= \mathbf{n}, \\ \mathbf{x}_2 - \mathbf{x}_1 &= \mathbf{p}_1(\mathbf{n}) \\ \mathbf{x}_3 - \mathbf{x}_1 &= \mathbf{p}_2(\mathbf{n}) \\ &\vdots \\ \mathbf{x}_{t+1} - \mathbf{x}_1 &= \mathbf{p}_t(\mathbf{n}) \end{aligned}$$

has monochromatic solutions $\{\mathbf{x}_0, \mathbf{x}_1, \cdots, \mathbf{x}_{t+1}\}$ with $\mathbf{n} = \mathbf{x}_0$ choosable from any large enough rectangle in $\mathbf{Z}^r$.

Proof. For any $L \in \mathbf{N}$, we will denote by $\mathcal{R}_L$ the set of rectangles

$$B = [M_1, N_1] \times [M_2, N_2] \times \cdots \times [M_r, N_r] \subset \mathbf{Z}^r$$

with $\min_{1 \leq i \leq r}(N_i - M_i) \geq L$. Renumbering the sets C_i if necessary, let $(C_i)_{i=1}^m$, where $m \leq s$, consist of those C_i for which $d^*(C_i) > 0$. For $1 \leq i \leq m$, we may via Lemma 7.2 let $(X_i, \mathcal{A}_i, \mu_i, \{T_{\mathbf{n}}^{(i)}\}_{\mathbf{n} \in \mathbf{Z}^r})$ be a measure preserving system and let $A_i \in \mathcal{A}_i$ be measurable sets with $\mu_i(A_i) = d^*(C_i)$ such that for any $l \in \mathbf{N}$ and $\mathbf{n}_1, \cdots, \mathbf{n}_l \in \mathbf{Z}^r$ one has

$$d^*\Big(C_i \cap (C_i - \mathbf{n}_1) \cap \cdots \cap (C_i - \mathbf{n}_l)\Big) \geq \mu_i\Big(A_i \cap T_{\mathbf{n}_1}^{(i)} A_i \cap \cdots \cap T_{\mathbf{n}_l}^{(i)} A_i\Big).$$

Let $X = X_1 \times \cdots \times X_m$, $\mathcal{A} = \mathcal{A}_1 \otimes \cdots \otimes \mathcal{A}_m$, $\mu = \mu_1 \times \cdots \times \mu_m$, $T_{\mathbf{n}} = T_{\mathbf{n}}^{(1)} \times \cdots \times T_{\mathbf{n}}^{(m)}$, $\mathbf{n} \in \mathbf{Z}^r$, and let $A = A_1 \times \cdots \times A_m$. According to Theorem 6.12, there exists $\epsilon > 0$ such that the set

$$R = \Big\{\mathbf{n} : \mu\Big(A \cap T_{\mathbf{p}_1(\mathbf{n})} A \cap \cdots \cap T_{\mathbf{p}_t(\mathbf{n})} A\Big) > \epsilon\Big\}$$

is syndetic in $\mathbf{Z}^r$, and therefore has positive uniform lower density. Letting $\delta > 0$ be less than the uniform lower density of R, we may choose L sufficiently large that:

a) For every $B \in \mathcal{R}_L$, $|B \cap R| > \delta |B|$.

b) For every $B \in \mathcal{R}_L$ we have

$$\left|\left(\bigcup_{i=1}^{m} C_i\right) \cap B\right| > (1-\delta)|B|.$$

It follows that for any $B \in \mathcal{R}_L$ one has $(B \cap R \cap C_i) \neq \emptyset$ for some i, $1 \leq i \leq m$. For $\mathbf{n}$ in this set one has

$$\begin{aligned} d^*\Big(C_i \cap (C_i - \mathbf{p}_1(\mathbf{n})) \cap \cdots \cap (C_i - \mathbf{p}_t(\mathbf{n}))\Big) &\geq \mu_i\Big(A_i \cap T^{(i)}_{\mathbf{p}_1(\mathbf{n})} A_i \cap \cdots \cap T^{(i)}_{\mathbf{p}_t(\mathbf{n})} A_i\Big) \\ &\geq \mu\Big(A \cap T_{\mathbf{p}_1(\mathbf{n})} A \cap \cdots \cap T_{\mathbf{p}_t(\mathbf{n})} A\Big) \\ &> \epsilon. \end{aligned}$$

□

We now, by a series of examples, show that the notions of syndeticity, IP^*_+, IP^*, E-IP^* and PE-IP^* are of strictly increasing exclusivity. As noted earlier the odd integers comprise a IP^*_+-set set which which is not an IP^* set. Hence there are three other cases to consider.

Example 7.5 *A syndetic set which is not IP^*_+.*

Following [F2, Definition 1.1], we say that a set $E \subset \mathbf{Z}$ is *thick* provided its complement is not syndetic. Equivalently, E is thick if it contains arbitrarily long intervals. The contrapositive of Proposition 6.11 asserts that every thick set contains an IP-set. On the other hand, any thick set obviously contains disjoint pairs of subsets which are themselves thick. These two facts imply that if E is any thick set we may find an IP-set $(n_\alpha)_{\alpha \in \mathcal{F}}$ in E such that the set $E \setminus \{n_\alpha : \alpha \in \mathcal{F}\}$ is still thick.

Let $(k_n)_{n=1}^\infty$ be an ordering of $\mathbf{Z}$. We will inductively construct a sequence of IP-sets $(\Gamma_n)_{n=1}^\infty$ such that the union

$$E = \bigcup_{n=1}^{\infty} (k_n + \Gamma_n) \tag{7.1}$$

contains no two consecutive integers. Choose Γ_1 having no two consecutive integers and with the property that the complement of Γ_1 is thick.

Suppose one is able to chose $\Gamma_1, \cdots, \Gamma_m$ with the property that the set

$$E_m = \bigcup_{n=1}^{m} (k_n + \Gamma_n)$$

contains no two consecutive elements and such that E_m^c is thick. We will demonstrate that one may choose Γ_{m+1} so as to carry these properties to the next step. Let $D_m = \big(E_m \cup (E_m - 1) \cup (E_m + 1)\big)$. Then $D_m^c - k_{m+1}$ is a thick set and hence contains an IP-set Γ_{m+1} which a) contains no two consecutive elements, and b) has the property that $\big((D_m^c - k_{m+1}) \setminus \Gamma_{m+1}\big) = \big(D_m^c \setminus (k_{m+1} + \Gamma_{m+1})\big)$ is thick, so that in particular $\big(E_m^c \setminus (k_{m+1}\Gamma_{m+1})\big)$ is thick.

By a) and the fact that $(k_{m+1} + \Gamma_{m+1}) \subset D_m^c$ and Γ_{m+1} contains no two consecutive elements, the set

$$E_{m+1} = \bigcup_{n=1}^{m+1} (k_n + \Gamma_n)$$

contains no two consecutive elements. By b), E^c_{m+1} is thick. Continuing in this fashion, the set E defined by (7.1) will contain no two consecutive integers. Hence E^c is syndetic. Suppose $l \in \mathbf{Z}$. For some $n \in \mathbf{N}$ we have $l = k_n$, hence $(l+\Gamma_n) \subset E$, which implies that $\Gamma_n \cap (E^c - l) = \emptyset$. It follows that $E^c - l$ is not IP*. Since l was arbitrary, E^c is not an IP$^*_+$-set .

□

We are indebted to N. Hindman for supplying the following example:

Example 7.6 *An IP*-set which is not E-IP*.*

Let $n_\alpha = \sum_{i\in\alpha} 2^{2i}$ and $m_\alpha = \sum_{i\in\alpha} 2^{2i+1}$ for $\alpha \in \mathcal{F}$. Every non-negative integer k has a unique representation $k = n_\alpha + m_\beta$, where $\alpha, \beta \in \mathcal{F}_\emptyset$. Let

$$E = \{n_\alpha + m_\beta : (\alpha, \beta) \in (\mathcal{F})^2_<\}.$$

E^c fails to be E-IP*. Indeed, there is no IP-ring $\mathcal{F}^{(1)}$ such that $n_\alpha + m_\beta$ lies in E^c for all $\alpha, \beta \in \mathcal{F}^{(1)}_\emptyset$. We will now show that E^c is IP* by demonstrating that E contains no IP-set.

Suppose that Γ is an IP-set and $\Gamma \subset E$. Let $k = (n_\alpha + m_\beta) \in \Gamma$ and let i be the largest member of β. Since Γ is an IP-set, the set $\{n \in \Gamma : (n+k) \in \Gamma\}$ contains an IP-set Γ'. Observe that any IP-set with finite range contains 0. (The only way an IP-set can fail to have infinite range is if all but finitely many of its generators are 0.) But $0 \notin E$. Hence $|\Gamma'| = \infty$ and hence Γ' contains some number $k' = n_{\alpha'} + m_{\beta'}$ which is a multiple of 2^{2i+2}. (We saw why in the proof of Theorem 0.9 appearing in Section 6.) $k+k'$ lies in Γ (therefore in E as well) and has the form $n_{\alpha\cup\alpha'} + m_{\beta\cup\beta'}$. By the uniqueness of the representation, this is a contradiction to the fact that $\alpha' > \beta$.

□

Example 7.7 *An E-IP*-set which is not PE-IP*.*

First, note that given any IP*-set R and any IP-set Γ, one may find an IP-set Γ' with $\Gamma' \subset (R \cap \Gamma)$. Consider now the set of squares $E = \{n^2 : n \in \mathbf{Z}\}$. We claim that no shift of E contains an infinite IP-set. To see this, consider that for any infinite IP-set Γ and non-zero $k \in \Gamma$ the set $\Gamma \cap (\Gamma - k)$ is infinite, while $E \cap (E-k)$ is finite for all k.

It follows that $E^c + k$ is IP* for every k. We claim that E^c is E-IP*. To see this, suppose that we are given k IP-sets $(n^{(1)}_\alpha)_{\alpha\in\mathcal{F}}, (n^{(2)}_\alpha)_{\alpha\in\mathcal{F}}, \cdots, (n^{(k)}_\alpha)_{\alpha\in\mathcal{F}}$. For simplicity, let us assume that $k = 2$. The ideas are the same for general k. Since E^c is IP*, by passing to a subring $\mathcal{F}^{(1)}$ we may assume that $n^{(1)}_\alpha \in E^c$ for all $\alpha \in \mathcal{F}^{(1)}$. Passing twice more to subrings we may assume that $n^{(2)}_\alpha \in E^c$ and that $(n^{(1)}_\alpha + n^{(2)}_\alpha) \in E^c$ for all $\alpha \in \mathcal{F}^{(1)}$. Fix $\alpha_1 \in \mathcal{F}^{(1)}$. Pass to a subring having indices $\alpha > \alpha_1$.

Since $E^c - n^{(1)}_{\alpha_1}$ is IP*, by passing to a subring we may assume that $n^{(1)}_\alpha \in E^c - n^{(1)}_{\alpha_1}$, that is $(n^{(1)}_\alpha + n^{(1)}_{\alpha_1}) \in E^c$, for all $\alpha \in \mathcal{F}^{(1)}$. Passing eight more times to subrings we may assume that $(n^{(1)}_\alpha + n^{(2)}_{\alpha_1}) \in E^c$, $(n^{(1)}_\alpha + n^{(1)}_{\alpha_1} + n^{(2)}_{\alpha_1}) \in E^c$, $(n^{(2)}_\alpha + n^{(1)}_{\alpha_1}) \in E^c$, $(n^{(2)}_\alpha + n^{(2)}_{\alpha_1}) \in E^c$, $(n^{(2)}_\alpha + n^{(1)}_{\alpha_1} + n^{(2)}_{\alpha_1}) \in E^c$, $(n^{(1)}_\alpha + n^{(2)}_\alpha + n^{(1)}_{\alpha_1}) \in E^c$, $(n^{(1)}_\alpha + n^{(2)}_\alpha + n^{(2)}_{\alpha_1}) \in E^c$, and $(n^{(1)}_\alpha + n^{(2)}_\alpha + n^{(1)}_{\alpha_1} + n^{(2)}_{\alpha_1}) \in E^c$ for all $\alpha \in \mathcal{F}^{(1)}$. Fix $\alpha_2 \in \mathcal{F}^{(1)}$ and pass to a subring with indices $\alpha > \alpha_2$.

Continue in this fashion. Having chosen $\alpha_1, \cdots, \alpha_t$, pass to a subring $\mathcal{F}^{(1)}$ having indices $\alpha > \alpha_t$ which has the property that $(n^{(1)}_\alpha + s) \in E^c$, $(n^{(2)}_\alpha + s) \in E^c$, and $(n_\alpha^{(1)} + n_\alpha^{(2)} + s) \in E^c$ for all $\alpha \in \mathcal{F}^{(1)}$ and all s in the set

$$\{n_\gamma^{(1)} + n_\eta^{(2)} : \gamma, \eta \in FU\{\alpha_1, \cdots, \alpha_t\}\}.$$

Let $\alpha_{t+1} \in \mathcal{F}^{(1)}$.

One now checks the IP-ring we have generated, namely $\mathcal{F}^{(2)} = FU\{\alpha_1, \alpha_2, \cdots\}$, has the property that for every $\alpha, \beta \in \mathcal{F}_\emptyset^{(2)}$ we have $(n_\alpha^{(1)} + n_\beta^{(2)}) \in E^c$. Taking for granted that this method works for general k, we have shown that E^c is E-IP*. Clearly E^c is not PE-IP*.

□

Our final pair of examples are strengthenings (in different directions), of Example 7.5.

Example 7.8 (*A set in* $\mathbf{Z}$ *of Banach lower density 1 which is not an* IP^*_+*-set* .)

Let E_0 be an IP-set in $\mathbf{N}$ with the property that in any interval of length 3^k, $k = 0, 1, 2, \cdots$, E_0 contains at most $2^k + 1$ points. ($E_0 = FS\big((3^n)_{n=0}^\infty\big)$ has this property.) E_0^c being thick, we may find an IP-set E_1 such that the distance from any point of E_1 to any point of E_0 is at least $1 + 3^1 = 4$. By then passing if necessary to a sub-IP-set, we may assume that E_1 has the form $E_1 = FS\big((a_n)_{n=0}^\infty\big)$, where $3 < a_0$ and $a_{i+1} > 3a_i$, $i = 0, 1, 2, \cdots$. Hence any interval of length 3^{k+1} contains at most $2^k + 1$ points of E_1. Having chosen $E_0, E_1, \cdots, E_{n-1}$ such that $F^c = \big(\bigcup_{i=0}^{n-1} i + E_i\big)^c$ is thick, choose an IP-set $E_n \subset \mathbf{N}$ every point of which is distance at least $n + 3^n$ from F and such that any interval of length 3^{k+n} contains at most $2^k + 1$ points of E_n. Continue the choosing.

Put $E = \bigcup_{n=0}^\infty (n + E_n)$. Let now $n \in \mathbf{N}$ and suppose I is an interval of length 3^n. Then I contains:

$$\begin{aligned}
&\text{at most 1 point from } \bigcup_{k=n}^\infty (k + E_k),\\
&\text{at most } 1 + 2^1 \text{ points from } (n-1) + E_{n-1},\\
&\text{at most } 1 + 2^2 \text{ points from } (n-2) + E_{n-2},\\
&\qquad\vdots\\
&\text{at most } 1 + 2^{n-1} \text{ points from } 1 + E_1, \text{ and}\\
&\text{at most } 1 + 2^n \text{ points from } E_0,
\end{aligned}$$

for a grand total of at most $n + 2^{n+1}$ points from E in I. Since $\frac{n+2^{n+1}}{3^n} \to 0$ as $n \to \infty$, this implies that E has Banach upper density 0 (it is sufficient to look at intervals of length 3^n in order to make this determination). Therefore $-E \cup E$ has Banach upper density 0 and contains, for every $n \in \mathbf{Z}$, a shift of an IP-set by n. This implies that no shift of $(-E \cup E)^c$ is IP*, that is, $(-E \cup E)^c$ is not a IP^*_+-set . On the other hand, $(-E \cup E)^c$ has Banach lower density 1.

□

Example 7.9 (*A syndetic IP-set in* $\mathbf{N}$ *which is not IP**.)

Although this example is included more or less as a curiosity, it will provide a natural segway into the brief discussion of topological recurrence with which we conclude this section. Let $(d_i)_{i=1}^{\infty}$ be defined as follows: $d_1 = 1$, and for $n \in \mathbf{N}$, $d_{2n} = 2d_{2n-1} + 1$ and $d_{2n+1} = 2d_{2n} - 1$. One may easily check that

$$d_{2n} = 2 + \sum_{k=1}^{2n-1} d_k \text{ and } d_{2n+1} = 1 + \sum_{k=1}^{2n} d_k, \; n \in \mathbf{N}. \tag{7.2}$$

For $k \in \mathbf{N}$, set $E_k = FS\big((d_i)_{i=k}^{\infty}\big)$. Put $E = E_1$. A consequence of (7.2) is that E_1 (in fact, E_k for all k) is a syndetic IP-set. We will show that E is not IP* by constructing an IP-set $(n_\alpha)_{\alpha \in \mathcal{F}} \subset \mathbf{N}$ disjoint from E.

We claim that for all $k \in \mathbf{N}$, $\mathbf{N} \subset E_k - E_k - E_k$. Indeed, letting $p \in \mathbf{N}$, we have

$$p = \Big(\sum_{n=k+1}^{k+p} d_{2n}\Big) - \Big(\sum_{n=k+1}^{k+p} d_{2n-1}\Big) - \Big(\sum_{n=k+1}^{k+p} d_{2n-1}\Big).$$

Consider now $1_E : \mathbf{N} \to \{0, 1\}$. We view 1_E as an infinite word on the alphabet $\{0, 1\}$ and claim that for each finite word w occuring in E, there exist natural numbers n and m such that w occurs in 1_E, beginning say at place m, at place n, and at place $m + n$. To see this, let us assume that a word w of length q occurs in 1_E beginning at place p. Fix k such that $d_k > p + q$. One easily checks (recall that E is an IP-set each of whose generators are greater than the sum of the previous ones), that w must occur beginning at every point of $E_k + p$. By the earlier claim, $p \in (E_k - E_k - E_k)$, which means that $(p + E_k) \bigcap \big((p + E_k) - (p + E_k)\big) = \big((p+E_k) \cap (E_k - E_k)\big)$ is non-empty. In other words, $\big((p+E_k)+(p+E_k)\big) \bigcap (p+E_k)$ is non-empty; that is, there exist m, n such that $\{m, n, m+n\} \subset p + E_k$. But w occurs beginning at each point of $p + E_k$, proving the claim (one may check that as a matter of fact, we are actually getting $m = n$ here, but this is not important for us).

We now inductively define a sequence $(n_k)_{k=1}^{\infty}$ such that $FS\big((n_k)_{k=1}^{\infty}\big) \subset E^c$. Observe that the single letter word $w_1 = 0$ occurs in 1_E, therefore by the claim just proved, there exist $m_1, n_1 \in \mathbf{N}$ such that w_1 occurs beginning at m_1, n_1 and $n_1 + m_1$. In other words, $\{n_1, m_1, n_1 + m_1\} \subset E^c$. Let w_2 be the $(n_1 + 1)$-letter word running from m_1 to $n_1 + m_1$ inclusive:

$$w_2 = \overbrace{0 * \ldots *}^{n_1} 0.$$

(This word begins and ends with 0; the $*$'s represent the letters in between the zeros, which can be 0's or 1's.) By the claim there exist n_2 and m_2 such that w_2 occurs beginning at n_2, m_2, and $n_2 + m_2$. One may check now that $\{n_1, n_2, n_2 + n_1\} \subset E^c$. Let w_3 be the $(n_1 + n_2 + 1)$-letter word that runs from m_2 to $m_2 + n_1 + n_2$ inclusive:

$$w_3 = \underbrace{\overbrace{0 * \ldots *}^{n_1} 0 * \ldots\ldots\ldots *}_{n_2} 0 * \ldots * 0$$

By the claim there exist n_3 and m_3 such that w_3 occurs beginning at n_3, m_3 and $n_3 + m_3$. Then $\big\{n_1, n_2, n_1 + n_2, n_3, n_3 + n_1, n_3 + n_2, n_3 + n_2 + n_1\big\} \subset E^c$. Continue and put $n_\alpha = \sum_{i \in \alpha} n_i$.

□

The constuction above provides us, as we shall presently exhibit, with an example of a *minimal* (i.e. having no non-trivial closed invariant sets) topological dynamical system (X,T), an open set U containing a point $x \in X$, and an IP-set $(n_\alpha)_{\alpha\in\mathcal{F}}$ such that for no $\alpha \in \mathcal{F}$ does one have $T^{n_\alpha}x \in U$, exhibiting explicitly a phenomenon which was shown in [F2] (see p. 183-184) to occur in all *topologically mixing* systems. (If X is compact and $T : X \to X$ continuous then the system (X,T) is called topologically mixing if for every pair of non-empty open sets A and B the set $\{n : (A \cap T^{-n}B) \neq \emptyset\}$ is thick.)

Here is the construction: let $\mathbf{N}_0 = \mathbf{N} \cup \{0\}$ and put $\Omega = \{0,1\}^{\mathbf{N}_0}$, giving Ω the product topology. Let $x = 1_{E\cup\{0\}} \in \Omega$, where E is the syndetic IP-set constructed in Example 7.9, and put $X = \overline{\{T^n x : n \in \mathbf{N}\}}$, where T is the shift $T\xi(n) = \xi(n+1)$. It is well known that X is minimal if and only if every word (that is, every finite sequence of 0's and 1's) that occurs in γ occurs along a syndetic set of starting places. But this is true (see the construction above, specifically the point at which it is pointed out that any word, if it occurs, occurs along a shift of the (syndetic) IP-set E_k). Therefore, (X,T) is a minimal system. (It is topologically mixing as well, but we do not show this.) Letting $U = \{\xi \in X : \xi(0) = 1\}$, we have $x \in U$, but clearly $T^{n_\alpha}x \notin U$ for all $\alpha \in \mathcal{F}$, where $(n_\alpha)_{\alpha\in\mathcal{F}}$ is the IP-set disjoint from E which was constructed in Example 7.9.

Thus we see that IP-sets are not so suitable for recurrence as to intersect non-trivially the set of returns of a point to one of its neighborhoods for (even minimal) topological systems. They are suitable enough, however, to intersect non-trivially the set of returns of any open set to itself in minimal topological systems, as has been previously noted (see for example [F2], p. 53, where this fact is implicit). As a matter of fact, it is known that for minimal systems, polynomial multiple recurrence along IP-sets occurs for non-empty open sets (see [BL1], [BL2]). may be viewed as a slight sharpening of the $\mathbf{Z}^k$ case of Lemma 1.10 from [BL2].

Theorem 7.10 Suppose that we have r commuting continuous maps $T_1, \cdots, T_r$ of a compact space $(X, \mathcal{B}, \mu)$ such that the system $(X, T_1, \cdots, T_r)$ is minimal. (That is, there are no non-trivial closed sets V with $V \subset T_i^{-1}V$, $1 \le i \le r$.) Let $k,t \in \mathbf{N}$, and assume that $p_{i,j}(x_1, \cdots, x_k) \in \mathbf{Q}[x_1, \cdots, x_k]$ with $p_{i,j}(\mathbf{N}_0^k) \subset \mathbf{N}_0$ and $p_{i,j}(0, \cdots, 0) = 0$, $1 \le i \le r$, $1 \le j \le t$. For every non-empty open set A, the set

$$R_A = \left\{ (n_1, \cdots, n_k) \in \mathbf{N}_0^k : \mu\Big(\bigcap_{j=1}^{t} \Big(\prod_{i=1}^{r} T_i^{p_{i,j}(n_1,\cdots,n_k)} \Big)^{-1} A \Big) > 0 \right\}$$

is a PE-IP*-set in $\mathbf{N}_0^k$.

Since it is well known that for any minimal system $(X, T_1, \cdots, T_r)$, there exists an invariant measure μ on X for which $\mu(U) > 0$ for every non-empty open set U, Theorem 7.10 is almost a corollary of Theorem 6.3. Not quite, however, for the continuous maps T_i appearing in Theorem 7.10 are allowed to be non-invertible, while Theorem 6.3 applies only to invertible systems. Not merely for this reason, but also to give a more complete picture of polynomial measure-theoretic recurrence along IP-sets, we will conclude this section by extending Theorem 6.3 to non-invertible systems (Theorem 7.12 below). We remark that Theorem 7.12 is several orders of magnitude deeper than Theorem 7.10, which can be inferred from the results in [BL2], where only combinatorial methods (in the guise of topological dynamics) are employed. Nevertheless, we do point out that in light of the existence of invariant measures, having obtained Theorem 7.12, Theorem 7.10 follows. First, we have a lemma.

Lemma 7.11 Suppose that we have r commuting (possibly non-invertible) measure preserving transformations $T_1, \cdots, T_r$ of a probability space $(X, \mathcal{B}, \mu)$, and let $A \in \mathcal{C}$. There exists a probability space $(Y, \mathcal{C}, \nu)$, commuting invertible measure preserving transformations $S_1, \cdots, S_r$ of Y, and a set $B \in \mathcal{C}$ such that for every $m \in \mathbf{N}$ and every $\mathbf{n}_1, \cdots, \mathbf{n}_m \in \mathbf{N}_0^r$,

$$\mu\big(A \cap T_{\mathbf{n}_1}^{-1} A \cap \cdots \cap T_{\mathbf{n}_m}^{-1} A\big) = \nu\big(B \cap S_{-\mathbf{n}_1} B \cap \cdots \cap S_{-\mathbf{n}_m} B\big). \tag{7.3}$$

Proof. Since the construction of the space Y is so similar to the construction of the space X in the proof of Theorem 7.2, we will only supply a sketch. According to the pointwise ergodic theorem, for almost every $x \in X$ we have

$$\lim_{N\to\infty} \frac{1}{N^r} \sum_{\mathbf{n}\in[1,\cdots,N]^r} 1_{A\cap T_{\mathbf{n}_1}^{-1} A \cap\cdots\cap T_{\mathbf{n}_m}^{-1} A}(T^n x) = \mu\big(A \cap T_{\mathbf{n}_1}^{-1} A \cap \cdots \cap T_{\mathbf{n}_m}^{-1} A\big) \tag{7.4}$$

for every $m \in \mathbf{N}$ and every choice of $\mathbf{n}_1, \cdots, \mathbf{n}_m \in \mathbf{N}^r$. Pick such an x and let $E = \{\mathbf{n} \in \mathbf{N}_0^r : T_{\mathbf{n}} x \in A\}$. Let $Y = \{0,1\}^{\mathbf{Z}^r}$ and let $S_1, \cdots, S_r$ be the coordinate shifts. Choose (see Theorem 7.2) an increasing sequence $(N_k)_{k=1}^{\infty} \subset \mathbf{N}$ such that

$$\lim_{k\to\infty} \frac{1}{N_k^r} \sum_{\mathbf{n}\in[1,\cdots,N_k]^r} f(S_{\mathbf{n}} x) = I(f)$$

exists for all $f \in \mathcal{C}(Y)$. Then I is a positive linear functional which is invariant under S_i, $1 \le i \le r$, hence is given by integration against an invariant Borel measure ν. Let $B = \{\gamma \in Y : \gamma(0) = 1\}$. One now derives (7.3) as a consequence of (7.4).

□

The extension of Theorem 6.3 to the non-invertible situation is, in light of Lemma 7.11, immediate.

Theorem 7.12 Suppose that we have r commuting (possibly non-invertible) measure preserving transformations $T_1, \cdots, T_r$ of a probability space $(X, \mathcal{B}, \mu)$. Let $k, t \in \mathbf{N}$, and assume that $p_{i,j}(x_1, \cdots, x_k) \in \mathbf{Q}[x_1, \cdots, x_k]$ with $p_{i,j}(\mathbf{N}_0^k) \subset \mathbf{N}_0$ and $p_{i,j}(0, \cdots, 0) = 0$, $1 \le i \le r$, $1 \le j \le t$. Then for every $A \in \mathcal{B}$ with $\mu(A) > 0$ the set

$$R_A = \left\{(n_1, \cdots, n_k) \in \mathbf{N}_0^k : \mu\Big(\bigcap_{j=1}^{t} \Big(\prod_{i=1}^{r} T_i^{p_{i,j}(n_1,\cdots,n_k)}\Big)^{-1} A\Big) > 0\right\}$$

is a PE-IP*-set in $\mathbf{N}_0^k$.

CHAPTER 8

FOR FUTURE INVESTIGATION

This short chapter is devoted a discussion of some natural open problems and conjectures which are suggested by the results of this paper. Two such problems (Questions 1 and 2) were posed in Chapter 6. The three we formulate here are of a somewhat weightier nature.

As stated in the introduction, there is a difference in scope between the *linear* IP multiple recurrence theorem of [FK2] and the *polynomial* IP multiple recurrence theorem (we'll take Theorem 0.9 as the model) proved above. Namely, the "linear" result is known to hold for infinitely generated IP-systems of operators. Indeed, let us formulate the theorem from [FK2] precisely.

Let $\{S_i\}_{i=1}^{\infty}$ be a countable sequence of commuting invertible measure preserving transformations of a probability space $(X, \mathcal{A}, \mu)$. Putting $T_\alpha = \prod_{i\in\alpha} S_i$, $\alpha \in \mathcal{F}$, the $\mathcal{F}$-sequence $\{T_\alpha\}_{\alpha\in\mathcal{F}}$ is called an IP-system of measure preserving transformations on X (notice that $T_{\alpha\cup\beta} = T_\alpha T_\beta$ whenever $\alpha \cap \beta = \emptyset$, and indeed this property together with commutativity of the T_α's is an alternative characterization for IP-systems). Two IP-systems $\{T_\alpha^{(1)}\}_{\alpha\in\mathcal{F}}$ and $\{T_\alpha^{(2)}\}_{\alpha\in\mathcal{F}}$ are said to commute if $T_\alpha^{(1)} T_\beta^{(2)} = T_\beta^{(2)} T_\alpha^{(1)}$ for all $\alpha, \beta \in \mathcal{F}$.

Theorem. ([FK2]) Let $(X, \mathcal{A}, \mu)$ be a probability space and let $k \in \mathbf{N}$. For any commuting IP-systems of measure preserving transformations $\{T_\alpha^{(i)}\}_{\alpha\in\mathcal{F}}$, $1 \le i \le k$, on X, and any $A \in \mathcal{A}$ with $\mu(A) > 0$, there exists $\alpha \in \mathcal{F}$ such that

$$\mu\Big(\bigcap_{i=1}^{k} \big(T_\alpha^{(i)}\big)^{-1} A\Big) > 0.$$

A full "polynomial-type" generalization of the above theorem can be formulated by defining a variant of IP-systems, as is done in the last section of [BFM]. Namely, we say that an $\mathcal{F}$-sequence $\{V_\alpha\}_{\alpha\in\mathcal{F}}$ in a multiplicative commutative group G with identity I is a *VIP*-system if for some d (called the *degree* of the system if it is the least such) we have

$$\prod_{0\le i_1<\cdots<i_k\le d} V_{\alpha_{i_1}\cup\cdots\cup\alpha_{i_k}}^{(-1)^k} = I, \text{ if } \alpha_i \cap \alpha_j = \emptyset,\ 0 \le i < j \le d.$$

For $d = 1$, this reduces to the alternative characterization of IP-systems noted parenthetically above. If G is a finitely generated group, any VIP-system in G is said to be finitely generated. One may check that members of $\mathbf{PE}_{\le d}(1) \setminus \mathbf{PE}_{\le (d-1)}(1)$ (see Definition 1.1 above) are finitely generated VIP-systems of degree d.

More typically, however, VIP-systems can have infinitely many generators. For example, let $\{S_{ij}\}_{i,j\in\mathbf{N}}$ be members of a commutative group and put $T_\alpha = \prod_{(i,j)\in\alpha\times\alpha} S_{ij}$, $\alpha\in\mathcal{F}$. Then $\{T_\alpha\}_{\alpha\in\mathcal{F}}$ is a VIP-system of degree (at most) 2.

We remark that the topological version of the following conjecture is known to be true ([BL2]).

Conjecture 1. If $k\in\mathbf{N}$, $(X,\mathcal{A},\mu)$ is a probability space and $\{V_\alpha^{(i)}\}_{\alpha\in\mathcal{F}}$, $1\le i\le k$ are commuting VIP-systems of measure preserving transformations of X then for every $A\in\mathcal{A}$ with $\mu(A)>0$ there exists $\alpha\in\mathcal{F}$ with

$$\mu\Big(\bigcap_{i=1}^{k}\big(V_\alpha^{(i)}\big)^{-1}A\Big)>0.$$

The first place at which the proof presented in this paper breaks down if one were to attempt to apply it to Conjecture 1 is in Theorem 2.17 (from [BFM]). Indeed, this theorem fails to transfer to general VIP-systems. A counterexample is given in Section 3 of [BFM]–a weakly convergent VIP-system (of degree 2) of unitary operators whose limit is not a projection. (Weak limits of IP-systems of unitary operators, on the other hand, must be projections, even if the systems are infinitely generated. See [FK2], Theorem 1.7.) As it happens, in many cases it is enough for recurrence purposes to know that a weak limit is positive.

Conjecture 2. If $\mathcal{H}$ is a Hilbert space, $\mathcal{F}^{(1)}$ is an IP-ring and $\{V_\alpha\}_{\alpha\in\mathcal{F}}$ is a VIP-system of unitary operators on $\mathcal{H}$ with

$$\operatorname*{IP-lim}_{\alpha\in\mathcal{F}^{(1)}} V_\alpha f = Qf$$

existing weakly for all $f\in\mathcal{H}$, then $\langle f, Qf\rangle\ge 0$ for all $f\in\mathcal{H}$.

Conjecture 2, if true, would already settle in the affirmative the $k=1$ case of Conjecture 1; that is, the case of single recurrence.

We will wrap things up by quoting a conjecture (from [B2], p. 56) that a "density polynomial Hales-Jewett theorem" holds which would extend both the partition results from [BL2] and the density version of the ("linear") Hales-Jewett theorem proved in [FK4]. For $q,d,N\in\mathbf{N}$ let $\mathcal{M}_{q,d,N}$ be the set of q-tuples of subsets of $\{1,2,\cdots,N\}^d$:

$$\mathcal{M}_{q,d,N}=\big\{(\alpha_1,\cdots,\alpha_q):\alpha_i\subset\{1,2,\cdots,N\}^d,\ i=1,2,\cdots,q\big\}.$$

Conjecture 3. For any $q,d\in\mathbf{N}$ and $\epsilon>0$ there exists $C=C(q,d,\epsilon)$ such that if $N>C$ and a set $S\subset\mathcal{M}_{q,d,N}$ satisfies $\frac{|S|}{|\mathcal{M}_{q,d,N}|}>\epsilon$ then S contains a "simplex" of the form:

$$\big\{(\alpha_1,\alpha_2,\cdots,\alpha_q),(\alpha_1\cup\gamma^d,\alpha_2,\cdots,\alpha_q),(\alpha_1,\alpha_2\cup\gamma^d,\cdots,\alpha_q),\\ \cdots,(\alpha_1,\alpha_2,\cdots,\alpha_q\cup\gamma^d)\big\},$$

where $\gamma\subset\mathbf{N}$ is a non-empty set and $\alpha_i\cap\gamma^d=\emptyset$ for all $i=1,2,\cdots,q$.

Conjecture 3 may well represent the ultimate plateau for this particular line of investigation.

APPENDIX

Multiparameter weakly mixing PET

In this appendix we give a very general weak mixing polynomial ergodic theorem cited in the proof of Theorem 6.4 and a corollary which is needed for Theorem 6.5.

Some notation follows. Let $k \in \mathbf{N}$. The Banach densities we have been using are actually generalizations of the more well-known notions of upper and lower densities defined by, for $E \subset \mathbf{Z}^k$,

$$\overline{d}(E) = \limsup_{N \to \infty} \frac{|E \cap \{-N, -N+1, \cdots, N\}^k|}{(2N+1)^k}$$

and

$$\underline{d}(E) = \liminf_{N \to \infty} \frac{|E \cap \{-N, -N+1, \cdots, N\}^k|}{(2N+1)^k}.$$

(If $\overline{d}(E) = \underline{d}(E)$ then we refer to this common value as the *density* of E.) Both the normal notion of density and the notion of Banach density may be used to define a mode a convergence for sequences indexed by $\mathbf{Z}^k$: if x is a real number and $(x_{\mathbf{n}})_{\mathbf{n} \in \mathbf{Z}^k}$ is a sequence of reals, let us write

$$UD - \lim_{\mathbf{n} \in \mathbf{Z}^k} x_{\mathbf{n}} = x \qquad \Big(\text{respectively } D - \lim_{\mathbf{n} \in \mathbf{Z}^k} x_{\mathbf{n}} = x\Big)$$

if for every $\epsilon > 0$ the set $\{\mathbf{n} \in \mathbf{Z}^k : |x_{\mathbf{n}} - x| > \epsilon\}$ has Banach lower density (respectively density) 0. (Equivalently, if for every $\epsilon > 0$ the set $\{\mathbf{n} \in \mathbf{Z}^k : |x_{\mathbf{n}} - x| < \epsilon\}$ has Banach upper density 1 (respectively density 1).)

Theorem A1. Suppose that $(X, \mathcal{B}, \mu)$ is a probability space and $\{T_{\mathbf{n}}\}_{\mathbf{n} \in \mathbf{Z}^r}$ is a totally weakly mixing measure preserving $\mathbf{Z}^r$-action generated by $T_1, \cdots, T_r$. Suppose that $t \in \mathbf{N}$ and $p_{i,j}(x_1, \cdots, x_k) \in \mathbf{Q}[x_1, \cdots, x_k]$ with $p_{i,j}(\mathbf{Z}^k) \subset \mathbf{Z}$, $1 \leq i \leq r$, $1 \leq j \leq t$ such that for any $1 \leq j_1 \neq j_2 \leq t$, the functions

$$\begin{aligned} &(l_1, \cdots, l_k) \\ \to &(p_{1,j_1}(l_1, \cdots, l_k) - p_{1,j_2}(l_1, \cdots, l_k), \cdots, p_{r,j_1}(l_1, \cdots, l_k) - p_{r,j_2}(l_1, \cdots, l_k)) \end{aligned}$$

are not constant. Suppose that $f_1, f_2, \cdots, f_t \in L^\infty(X, \mathcal{B}, \mu)$. Then

$$UD - \lim_{\mathbf{n} \in \mathbf{Z}^k} \Big(\int \prod_{j=1}^{t} \Big(\prod_{i=1}^{r} T_i^{p_{i,j}(n_1, \cdots, n_k)} \Big) f_j \, d\mu \Big) = \prod_{j=1}^{t} \Big(\int f_j \, d\mu \Big).$$

Before proving Theorem A1 we will introduce some notation and prove a few preparatory lemmas. Much of what we do here is completely analogous to what we do in Section 4, the primary differences being present to accomodate a slightly different kind of polynomial expression. The polynomial expressions we are dealing with here are of the type

$$T(\mathbf{n}) = T(n_1, \cdots, n_k) = \prod_{i=1}^{r} T_i^{p_i(n_1, \cdots, n_k)}. \tag{A.1}$$

Let us define the weight $w(T)$ of the expression T as given by (A.1) to be the ordered pair (t, d), where t is the maximum index i such that $\deg p_i > 0$ and $d = \deg p_t$. (Recall the mneumonic device introduced before Example 4.3: "(last T, last degree)".) If $\deg p_i = 0$ for all i then we write $w(T) = (0, 0)$.

Example A2. $w(T_1^{n_1 n_2 - 3} T_2^{n_2^9 - 6} T_3^{n_1^3 n_2^4 + n_3^5}) = (3, 7)$, $w(T_5^{n_3^4} T_8^{17}) = (5, 4)$.

The weights will be ordered as before, namely we write $(t, d) < (h, k)$ if either $t < h$ or if $t = h$ and $d < k$. Analogous to the situation with the polynomial expressions in the text, if

$$T(\mathbf{n}) = \prod_{i=1}^{r} T_i^{p_i(n_1, \cdots, n_k)} \text{ and } S(\mathbf{n}) = \prod_{i=1}^{r} T_i^{q_i(n_1, \cdots, n_k)}$$

with $w(S) = w(T) = (t, d)$ then we write $S \sim T$ if p_t and q_t are identical in their dth degree terms. $\sim$ is an equivalence relation.

Example A3. $T_1^{n_5} T_4^{n_2 + n_3 - 2} T_5^{14} \sim T_2^{n_2^3 + 12} T_4^{n_2 + n_3 + 7} T_6^{13} \not\sim T_6^{13}$.

For a polynomial of k variables $p(n_1, \cdots, n_k)$, write

$$p^{(2)}(\mathbf{n}, \mathbf{h}) = p^{(2)}(n_1, \cdots, n_k, h_1, \cdots, h_k) = p(\mathbf{n} + \mathbf{h}) - p(\mathbf{n}) - p(\mathbf{h}).$$

(Compare Definition 2.5). For a polynomial expression $T(\mathbf{n}) = \prod_{i=1}^{r} T_i^{p_i(\mathbf{n})}$ write

$$T^{(2)}(\mathbf{n}, \mathbf{h}) = T(\mathbf{n} + \mathbf{h})\big(T(\mathbf{n})T(\mathbf{h})\big)^{-1} = \prod_{i=1}^{r} T_i^{p_i^{(2)}(\mathbf{n}, \mathbf{h})}.$$

Example A4. Suppose $p(\mathbf{n}) = p(n_1, n_2) = n_1 n_2^2$. Then

$$p^{(2)}(\mathbf{n}, \mathbf{h}) = p^{(2)}(n_1, n_2, h_1, h_2) = h_1 n_2^2 + 2n_1 n_2 h_2 + 2n_2 h_1 h_2 + n_1 h_2^2.$$

Suppose that A is a finite set of polynomial expressions of the form $(A.1)$, and suppose that the highest degree polynomial appearing in any of the expressions of A is of degree d. For each weight (a, b), $1 \le a \le r$, $1 \le b \le d$, let m_{ab} be the number of equivalence classes under $\sim$ represented in A by expressions of weight (a, b). The $r \times d$ matrix (m_{ab}) will be called the *weight matrix* of A. An $r \times d$ weight matrix (m_{ab}) is said to precede another $r \times d$ weight matrix (n_{ab}) if there exists a weight (i, j) such that $m_{ij} < n_{ij}$ and $m_{ab} = n_{ab}$ whenever $(a, b) > (i, j)$. This notion of "precedes" defines a well-ordering on the set of $r \times d$ weight matrices. It is by induction on the weight matrix of A under this ordering that Theorem A1 will be proved (PET-induction).

The following remarks will be used in the proof of Theorem A1.

Remarks A5.

(i) If $w(S) < w(T)$ then $ST \sim T$.

(ii) If $\deg p(\mathbf{n}) > 0$ and $\mathbf{h} \in \mathbf{Z}^k$ then the degree of $p^{(2)}(\mathbf{n}, \mathbf{h})$ as a polynomial in $\mathbf{n}$ is $\deg p - 1$.

(iii) If $w(T) > (0,0)$ then the weight of $T^{(2)}(\mathbf{n}, \mathbf{h})$ as an expression in $\mathbf{n}$ is less than the weight of T.

(iv) For fixed $\mathbf{h} \in \mathbf{Z}^k$, $T(\mathbf{n} + \mathbf{h}) \sim T(\mathbf{n})$.

(v) If $p(\mathbf{n})$ and $q(\mathbf{n})$ are polynomials then either (a) $p(\mathbf{n} + \mathbf{h}) + q(\mathbf{n})$ depends only on $\mathbf{h}$, in which case $p(\mathbf{n}) + q(\mathbf{n})$ is constant and $\deg p \leq 1$, or (b) the set of $\mathbf{h} \in \mathbf{Z}^k$ for which $p(\mathbf{n}+\mathbf{h})+q(\mathbf{n})$ depends only on $\mathbf{h}$ is of zero Banach upper density.

(vi) If $T(\mathbf{n})$ and $S(\mathbf{n})$ are polynomial expressions then either (a) $T(\mathbf{n}+\mathbf{h})S(\mathbf{n})$ depends only on $\mathbf{h}$, in which case $T(\mathbf{n})S(\mathbf{n})$ is constant and T is built from polynomials of degree ≤ 1, or (b) the set of $\mathbf{h} \in \mathbf{Z}^k$ for which $T(\mathbf{n} + \mathbf{h})S(\mathbf{n})$ depends only on $\mathbf{h}$ is of zero Banach upper density.

(vii) If $T \not\sim S_1 \sim S_2$ then $T^{-1}S_1 \sim T^{-1}S_2$.

Concerning the foregoing remarks, (i) and (ii) are routinely verified, (iii) is a consequence of (ii), (iv) is a consequence of (i) and (iii), and (vi) is a consequence of (v). As for (v), the set of $\mathbf{h}$ mentioned there is actually finite, indeed can contain at most 1 element. This however requires a proof. Zero density of the set is sufficient for our purposes and is an easy consequence of the fact that for any non-zero, k-variable polynomial $p(\mathbf{n})$ and any $t \in \mathbf{Z}$ the set $\{\mathbf{n} \in \mathbf{Z}^k : p(\mathbf{n}) = t\}$ is a zero density set. (The reason this is sufficient: if $p^{(2)}(\mathbf{n}, \mathbf{h})$ is non-zero then it is in effect a non-zero polynomial in $n_1, \cdots, n_k$ whose coefficients are non-zero polynomials in $\mathbf{h}$. It is only for those $\mathbf{h}$ at which these "coefficient polynomials" are equal to the negative of the corresponding (fixed) coefficients in $p(\mathbf{n}) + q(\mathbf{n})$ that there is any hope of $p(\mathbf{n} + \mathbf{h}) + q(\mathbf{n}) = p(\mathbf{n}) + p(\mathbf{h}) + p^{(2)}(\mathbf{n}, \mathbf{h}) + q(\mathbf{n})$ depending only on $\mathbf{h}$.)

We come now to the first of our lemmas. It is inspired by van der Corput's difference theorem. (See also Theorem 2.18 above, [B1] and [BM].)

Lemma A6. Suppose that $k \in \mathbf{N}$ and $(x_\mathbf{n})_{\mathbf{n}\in\mathbf{Z}^k}$ is a bounded sequence of vectors in a Hilbert space. If

$$D-\lim_{\mathbf{h}\in\mathbf{Z}^k}\left(\lim_{\substack{N_i-M_i\to\infty\\1\leq i\leq k}}\frac{1}{\prod_{i=1}^k(N_i-M_i)}\sum_{\mathbf{n}\in\prod_{i=1}^k[M_i+1,N_i]}\langle x_\mathbf{n}, x_{\mathbf{n}+\mathbf{h}}\rangle\right)=0$$

then

$$\lim_{\substack{N_i-M_i\to\infty\\1\leq i\leq k}}\left|\left|\frac{1}{\prod_{i=1}^k(N_i-M_i)}\sum_{\mathbf{n}\in\prod_{i=1}^k[M_i+1,N_i]}x_\mathbf{n}\right|\right|=0.$$

Proof. We begin with the observation that for any bounded sequence $(a_\mathbf{h})_{\mathbf{h}\in\mathbf{Z}^k}$ of real numbers with

$$D-\lim_{\mathbf{h}\in\mathbf{Z}^k} a_\mathbf{h} = 0,$$

we have

$$\lim_{H\to\infty}\sum_{\mathbf{h}\in[-2H,2H]^k}\frac{\prod_{i=1}^k(2H+1-|h_i|)}{(2H+1)^{2k}}a_\mathbf{h}=0,$$

where $\mathbf{h} = (h_1, \cdots, h_k)$. In order to see, this, first note that (we don't prove it),

$$\sum_{\mathbf{h}\in[-2H,2H]^k}\frac{\prod_{i=1}^k(2H+1-|h_i|)}{(2H+1)^{2k}}=1,$$

i.e. this is an averaging scheme. Let $\delta > 0$ be arbitrary and put $E = \{\mathbf{h} \in \mathbf{Z}^k : |a_{\mathbf{h}}| > \delta\}$. Then $d(E) = 0$, so there exists $k \in \mathbf{N}$ such that for all $H > k$ we have

$$\frac{|E \cap [-2H, 2H]^k|}{(4H+1)^k} \leq \frac{\delta}{2^k}.$$

Then

$$\begin{aligned}
&\sum_{\mathbf{h}\in[-2H,2H]^k} \frac{\prod_{i=1}^k (2H+1-|h_i|)}{(2H+1)^{2k}} a_{\mathbf{h}} \\
&= \sum_{\mathbf{h}\in[-2H,2H]^k\cap E} \frac{\prod_{i=1}^k (2H+1-|h_i|)}{(2H+1)^{2k}} a_{\mathbf{h}} + \sum_{\mathbf{h}\in[-2H,2H]^k\setminus E} \frac{\prod_{i=1}^k (2H+1-|h_i|)}{(2H+1)^{2k}} a_{\mathbf{h}} \\
&\leq \frac{\delta(2H+1)^k(4H+1)^k}{2^k(2H+1)^{2k}} + \delta < 2\delta.
\end{aligned}$$

This establishes the claim. Let now $\epsilon > 0$ be arbitrary and let $H \in \mathbf{N}$ be large enough that

$$\sum_{\mathbf{h}\in[-2H,2H]^k} \frac{\prod_{i=1}^k (2H+1-|h_i|)}{(2H+1)^{2k}} \lim_{\substack{N_i - M_i \to\infty \\ 1\leq i\leq k}} \frac{1}{\prod_{i=1}^k (N_i - M_i)} \sum_{\mathbf{n}\in\prod_{i=1}^k [M_i+1,N_i]} \langle x_{\mathbf{n}}, x_{\mathbf{n}+\mathbf{h}}\rangle < \epsilon,$$

where again $\mathbf{h} = (h_1, h_2, \cdots, h_k)$. We have

$$\lim_{\substack{N_i - M_i \to\infty \\ 1\leq i\leq k}} \left|\left| \frac{1}{\prod_{i=1}^k (N_i - M_i)} \sum_{\mathbf{n}\in\prod_{i=1}^k [M_i+1,N_i]} x_{\mathbf{n}} - \frac{1}{\prod_{i=1}^k (N_i - M_i)} \sum_{\mathbf{n}\in\prod_{i=1}^k [M_i+1,N_i]} \frac{1}{(2H+1)^k} \sum_{\mathbf{h}\in[-H,H]^k} x_{\mathbf{n}+\mathbf{h}} \right|\right| = 0.$$

This fact, together with the fact that ϵ is arbitrary, implies that we will be done if we can show that

$$\limsup_{\substack{N_i - M_i \to\infty \\ 1\leq i\leq r}} \left|\left| \frac{1}{\prod_{i=1}^k (N_i - M_i)} \sum_{\mathbf{n}\in\prod_{i=1}^k [M_i+1,N_i]} \frac{1}{(2H+1)^k} \sum_{\mathbf{h}\in[-H,H]^k} x_{\mathbf{n}+\mathbf{h}} \right|\right| \leq \epsilon.$$

But

$$\begin{aligned}
&\limsup_{\substack{N_i - M_i \to\infty \\ 1\leq i\leq r}} \left|\left| \frac{1}{\prod_{i=1}^k (N_i - M_i)} \sum_{\mathbf{n}\in\prod_{i=1}^k [M_i+1,N_i]} \frac{1}{(2H+1)^k} \sum_{\mathbf{h}\in[-H,H]^k} x_{\mathbf{n}+\mathbf{h}} \right|\right|^2 \\
&\leq \limsup_{\substack{N_i - M_i \to\infty \\ 1\leq i\leq r}} \frac{1}{\prod_{i=1}^k (N_i - M_i)} \sum_{\mathbf{n}\in\prod_{i=1}^k [M_i+1,N_i]} \left|\left| \frac{1}{(2H+1)^k} \sum_{\mathbf{h}\in[-H,H]^k} x_{\mathbf{n}+\mathbf{h}} \right|\right|^2 \\
&= \limsup_{\substack{N_i - M_i \to\infty \\ 1\leq i\leq r}} \frac{1}{\prod_{i=1}^k (N_i - M_i)} \sum_{\mathbf{n}\in\prod_{i=1}^k [M_i+1,N_i]} \frac{1}{(2H+1)^{2k}} \sum_{\mathbf{l},\mathbf{q}\in[-H,H]^k} \langle x_{\mathbf{n}+\mathbf{l}}, x_{\mathbf{n}+\mathbf{q}}\rangle.
\end{aligned}$$

We now plan to rewrite $\langle x_{\mathbf{n}+\mathbf{l}}, x_{\mathbf{n}+\mathbf{q}} \rangle$ as $\langle x_{\mathbf{n}}, x_{\mathbf{n}+\mathbf{h}} \rangle$, noting that as $N_i - M_i \to \infty$, $1 \leq i \leq k$, the typical vector $\mathbf{n} \in \prod_{i=1}^k [M_i + 1, N_i]$ is represented as $\mathbf{n} + \mathbf{l}$ in the above expression $(2H+1)^k$ times, and it is paired with the vector $\mathbf{n} + \mathbf{h}$ in the inner product zero times if $\mathbf{h} \notin [-2H, 2H]^k$ and $\prod_{i=1}^k \left(2H + 1 - |h_i|\right)$ times if $\mathbf{h} \in [-2H, 2H]^k$. (This is the number of solutions $\{l, q\}$ to the equation $\mathbf{h} = l - q$ for $l, q \in [-H, H]^k$. Again, we don't prove this.) So, continuing from the last line of the previous display,

$$= \limsup_{\substack{N_i - M_i \to \infty \\ 1 \leq i \leq r}} \frac{1}{\prod_{i=1}^k (N_i - M_i)} \sum_{\mathbf{n} \in \prod_{i=1}^k [M_i+1, N_i]} \frac{1}{(2H+1)^{2k}} \sum_{\mathbf{h} \in [-2H, 2H]^k} \prod_{i=1}^k \left(2H + 1 - |h_i|\right) \langle x_{\mathbf{n}}, x_{\mathbf{n}+\mathbf{h}} \rangle$$

$$= \sum_{\mathbf{h} \in [-2H, 2H]^k} \frac{\prod_{i=1}^k \left(2H + 1 - |h_i|\right)}{(2H+1)^{2k}} \lim_{\substack{N_i - M_i \to \infty \\ 1 \leq i \leq k}} \frac{1}{\prod_{i=1}^k (N_i - M_i)} \sum_{\mathbf{n} \in \prod_{i=1}^k [M_i+1, N_i]} \langle x_{\mathbf{n}}, x_{\mathbf{n}+\mathbf{h}} \rangle \leq \epsilon.$$

$\square$

The mean ergodic theorem will be used to help establish the initial case of the induction. Recall that a measure preserving system $(X, \mathcal{B}, \mu, \{T_{\mathbf{n}}\}_{\mathbf{n} \in \mathbf{Z}^r})$ is called ergodic if the only measurable invariant sets E (that is, $E \in \mathcal{B}$ with $\mu(E \triangle T_{\mathbf{n}} E) = 0$ for all $\mathbf{n} \in \mathbf{Z}^k$) satisfy $\mu(E) \in \{0, 1\}$. The following result is classical and well-known, so we omit the proof.

Lemma A7. (Mean ergodic theorem.) Suppose that $(X, \mathcal{B}, \mu, \{T_{\mathbf{n}}\}_{\mathbf{n} \in \mathbf{Z}^r})$ is a measure preserving system. Then

$$Pf = \lim_{\substack{N_i - M_i \to \infty \\ 1 \leq i \leq k}} \frac{1}{\prod_{i=1}^k (N_i - M_i)} \sum_{\mathbf{n} \in \prod_{i=1}^k [M_i+1, N_i]} T^{\mathbf{n}} f$$

exists in $L^2(X, \mathcal{B}, \mu)$ for all $f \in L^2(X, \mathcal{B}, \mu)$. Moreover, P is the orthogonal projection onto the space of $\{T_{\mathbf{n}}\}_{\mathbf{n} \in \mathbf{Z}^r}$-invariant functions.

Lemma A8. If $(a_{\mathbf{n}})_{\mathbf{n} \in \mathbf{Z}^r}$ is a sequence of reals and α is a real number such that

$$\lim_{\substack{N_i - M_i \to \infty \\ 1 \leq i \leq k}} \frac{1}{\prod_{i=1}^k (N_i - M_i)} \sum_{\mathbf{n} \in \prod_{i=1}^k [M_i+1, N_i]} a_{\mathbf{n}} = \alpha$$

and

$$\lim_{\substack{N_i - M_i \to \infty \\ 1 \leq i \leq k}} \frac{1}{\prod_{i=1}^k (N_i - M_i)} \sum_{\mathbf{n} \in \prod_{i=1}^k [M_i+1, N_i]} a_{\mathbf{n}}^2 = \alpha^2$$

then

$$UD - \lim_{\mathbf{n} \in \mathbf{Z}^r} a_{\mathbf{n}} = \alpha.$$

Proof. We have

$$\begin{aligned}
&\lim_{\substack{N_i-M_i\to\infty\\ 1\le i\le k}} \frac{1}{\prod_{i=1}^k (N_i-M_i)} \sum_{\mathbf{n}\in\prod_{i=1}^k [M_i+1,N_i]} (a_{\mathbf{n}}-\alpha)^2 \\
=&\lim_{\substack{N_i-M_i\to\infty\\ 1\le i\le k}} \frac{1}{\prod_{i=1}^k (N_i-M_i)} \sum_{\mathbf{n}\in\prod_{i=1}^k [M_i+1,N_i]} a_{\mathbf{n}}^2 \\
&-2\alpha\Big(\lim_{\substack{N_i-M_i\to\infty\\ 1\le i\le k}} \frac{1}{\prod_{i=1}^k (N_i-M_i)} \sum_{\mathbf{n}\in\prod_{i=1}^k [M_i+1,N_i]} a_{\mathbf{n}}\Big)+\alpha^2 \\
=&\alpha^2-2\alpha^2+\alpha^2=0.
\end{aligned}$$

On the other hand for any $\epsilon>0$,

$$\limsup_{\substack{N_i-M_i\to\infty\\ 1\le i\le r}} \frac{1}{\prod_{i=1}^k (N_i-M_i)} \sum_{\mathbf{n}\in\prod_{i=1}^k [M_i+1,N_i]} (a_{\mathbf{n}}-\alpha)^2 \ge \epsilon^2\Big(d^*\big(\{\mathbf{n}:|a_{\mathbf{n}}-\alpha|\ge\epsilon\}\big)\Big).$$

□

Lemma A9. If $k\in\mathbf{N}$, $l_1,\cdots,l_k\in\mathbf{Z}$ (not all zero), and $(X,\mathcal{B},\mu,T)$ is a weakly mixing system then for any $f,g\in L^\infty(X,\mathcal{B},\mu)$ we have

$$UD-\lim_{\mathbf{n}\in\mathbf{Z}^k}\int gT^{l_1n_1+\cdots+l_kn_k}f=\Big(\int f\,d\mu\Big)\Big(\int g\,d\mu\Big).$$

Proof. Let $a_{\mathbf{n}}=\int gT^{l_1n_1+\cdots+l_kn_k}f\,d\mu$, $\mathbf{n}=(n_1,n_2,\cdots,n_k)\in\mathbf{Z}^k$. We have

$$\begin{aligned}
&\lim_{\substack{N_i-M_i\to\infty\\ 1\le i\le k}} \frac{1}{\prod_{i=1}^k (N_i-M_i)} \sum_{\mathbf{n}\in\prod_{i=1}^k [M_i+1,N_i]} a_{\mathbf{n}} \\
=&\lim_{\substack{N_i-M_i\to\infty\\ 1\le i\le k}} \frac{1}{\prod_{i=1}^k (N_i-M_i)} \sum_{\mathbf{n}\in\prod_{i=1}^k [M_i+1,N_i]} \int gT^{l_1n_1+\cdots+l_kn_k}f\,d\mu \\
=&\int g\Big(\lim_{\substack{N_i-M_i\to\infty\\ 1\le i\le k}} \frac{1}{\prod_{i=1}^k (N_i-M_i)} \sum_{\mathbf{n}\in\prod_{i=1}^k [M_i+1,N_i]} T^{l_1n_1+\cdots+l_kn_k}f\Big)\,d\mu \\
=&\Big(\int g\,d\mu\Big)\Big(\int f\,d\mu\Big).
\end{aligned} \tag{A.2}$$

(The last equality in this display is by Theorem A7. The quantity in parentheses in the second to last line is equal to the projection of f onto the space of functions invariant under $T^{l_1},T^{l_2},\cdots,T^{l_k}$. In other words, this is equal to $\int f\,d\mu$.)

Since T is weakly mixing, $(X\times X,\mathcal{B}\otimes\mathcal{B},\mu\times\mu,T\times T)$ is an ergodic system.

Write as always $h \otimes k(x,y)$ for $h(x)k(y)$. Now

$$\begin{aligned}
&\lim_{\substack{N_i - M_i \to \infty \\ 1 \le i \le k}} \frac{1}{\prod_{i=1}^k (N_i - M_i)} \sum_{\mathbf{n} \in \prod_{i=1}^k [M_i+1, N_i]} a_{\mathbf{n}}^2 \\
= &\lim_{\substack{N_i - M_i \to \infty \\ 1 \le i \le k}} \frac{1}{\prod_{i=1}^k (N_i - M_i)} \sum_{\mathbf{n} \in \prod_{i=1}^k [M_i+1, N_i]} \Big(\int g T^{l_1 n_1 + \cdots + l_k n_k} f \, d\mu \Big)^2 \\
= &\lim_{\substack{N_i - M_i \to \infty \\ 1 \le i \le k}} \frac{1}{\prod_{i=1}^k (N_i - M_i)} \sum_{\mathbf{n} \in \prod_{i=1}^k [M_i+1, N_i]} \\
&\qquad \int (g \otimes g)(T \times T)^{l_1 n_1 + \cdots + l_k n_k} (f \otimes f) \, d\mu \times \mu \\
= &\int (g \otimes g) \Big(\lim_{\substack{N_i - M_i \to \infty \\ 1 \le i \le k}} \frac{1}{\prod_{i=1}^k (N_i - M_i)} \\
&\qquad \sum_{\mathbf{n} \in \prod_{i=1}^k [M_i+1, N_i]} (T \times T)^{l_1 n_1 + \cdots + l_k n_k} (f \otimes f) \Big) d\mu \times \mu \\
= &\Big(\int g \otimes g \, d\mu \times \mu \Big) \Big(\int f \otimes f \, d\mu \times \mu \Big) \\
= &\Big(\int g \, d\mu \Big)^2 \Big(\int f \, d\mu \Big)^2 .
\end{aligned}$$

Together with (A.2) and Lemma A8 this implies that

$$UD - \lim_{\mathbf{n} \in \mathbf{Z}^k} a_{\mathbf{n}} = UD - \lim_{n \in \mathbf{Z}} \int g T^{l_1 n_1 + \cdots + l_k n_k} f \, d\mu = \Big(\int g \, d\mu \Big) \Big(\int f \, d\mu \Big).$$

□

Proof of Theorem A1. Let us fix $r, d \in \mathbf{N}$ and proceed to prove Theorem A1 for a fixed totally weakly mixing $\mathbf{Z}^r$-action generated by $T_1, \cdots, T_r$ and polynomials p_{ij} of degree at most d. Since d is arbitrary this is sufficient.

Let $S_j(\mathbf{n}) = S_j(n_1, \cdots, n_k) = \prod_{i=1}^r T_i^{p_{i,j}(n_1, \cdots, n_k)}$ and set $A = \{S_1, \cdots, S_t\}$. The proof will be by induction on the weight matrix of A. By using certain identities, we may assume without loss of generality that $\int f_{i_0} \, d\mu = 0$ for some i_0, $1 \le i_0 \le t$ (as in the proof of Theorem 4.10). Hence what we are trying to establish is that

$$UD - \lim_{\mathbf{n} \in \mathbf{Z}^k} \Big(\int \prod_{j=1}^t \Big(\prod_{i=1}^r T_i^{p_{i,j}(n_1, \cdots, n_k)} \Big) f_j \, d\mu \Big) = 0. \tag{A.3}$$

The first non-trivial case is when A has weight matrix

$$\begin{pmatrix} 1 & 0 & \cdots & 0 \\ 0 & 0 & \cdots & 0 \\ \vdots & \vdots & \ddots & \vdots \\ 0 & 0 & \cdots & 0 \end{pmatrix}.$$

In this case A consists of a single polynomial expression which is linear in T_1 and constant in $T_2, \cdots, T_r$, and (possibly) a constant polynomial expression. (There is only one linear expression because any two such would, if in the same equivalence class under $\sim$, differ by a constant.) Since the case in which the constant polynomial expression doesn't occur is trivial, we will assume that it does occur. By possibly replacing f_1 and f_2 by appropriate images of these functions under the $\mathbf{Z}^r$-action, we may assume that the constant expression is the identity and the linear expression is given by $T_1^{l_1n_1+l_2n_2+\cdots+l_kn_k}$. Hence this case reduces to showing that

$$UD-\lim_{\mathbf{n}\in\mathbf{Z}^k}\left(\int f_1T_1^{l_1n_1+l_2n_2+\cdots+l_kn_k}f_2\,d\mu\right)=0$$

given that either $\int f_1\,d\mu=0$ or $\int f_2\,d\mu=0$. This is a consequence of Lemma A8 and establishes the initial case.

Let us assume, then, that the conclusion of Theorem A1 holds for every family of polynomial expressions whose weight matrix precedes that of A, and proceed to establish (A.3). Our first step is to show that

$$\lim_{\substack{N_i-M_i\to\infty\\1\le i\le k}}\left|\left|\frac{1}{\prod_{i=1}^k(N_i-M_i)}\sum_{\mathbf{n}\in\prod_{i=1}^k[M_i+1,N_i]}\prod_{j=1}^t S_j(\mathbf{n})f_j\right|\right|=0 \qquad (A.4)$$

under the additional restriction that A contains no constant expressions. (Notice that the addition or removal of a constant expression does not change the weight matrix of A.)

We will use Lemma A6. Let $x_{\mathbf{n}}=\prod_{j=1}^t S_j(\mathbf{n})f_j$. Take $L^2(X,\mathcal{B},\mu)$ to consist of real-valued functions only. For any $\mathbf{h}\in\mathbf{Z}^k$,

$$\begin{aligned}&\lim_{\substack{N_i-M_i\to\infty\\1\le i\le k}}\frac{1}{\prod_{i=1}^k(N_i-M_i)}\sum_{\mathbf{n}\in\prod_{i=1}^k[M_i+1,N_i]}\langle x_{\mathbf{n}},x_{\mathbf{n}+\mathbf{h}}\rangle\\ =&\lim_{\substack{N_i-M_i\to\infty\\1\le i\le k}}\frac{1}{\prod_{i=1}^k(N_i-M_i)}\sum_{\mathbf{n}\in\prod_{i=1}^k[M_i+1,N_i]}\qquad (A.5)\\ &\int\Big(\prod_{j=1}^t S_j(\mathbf{n})f_j\Big)\Big(\prod_{j=1}^t S_j(\mathbf{n}+\mathbf{h})f_j\Big)\,d\mu.\end{aligned}$$

Reindexing if necessary, let us assume that S_1 is of the minimal weight occuring in A.

For $\mathbf{h}\in\mathbf{Z}^k$, consider the family of polynomial expressions

$$\begin{aligned}A_{\mathbf{h}}=&\{S_j(\mathbf{n})S_1^{-1}(\mathbf{n}),S_j(\mathbf{n}+\mathbf{h})S_1^{-1}(\mathbf{n}):1\le j\le t\}\\ =&\{S_j(\mathbf{n})S_1^{-1}(\mathbf{n}),S_j(\mathbf{n})S_j^{(2)}(\mathbf{n},\mathbf{h})S_j(\mathbf{h})S_1^{-1}(\mathbf{n}):1\le j\le t\}.\end{aligned}$$

We claim that the weight matrix of $A_{\mathbf{h}}$ precedes that of A.

To see this, notice first of all that, by Remarks A5 (iv) and (vii), $S_j(\mathbf{n})S_1^{-1}(\mathbf{n})\sim S_j(\mathbf{n}+\mathbf{h})S_1^{-1}(\mathbf{n})$, $1\le j\le t$. This implies that the weight matrix of $A_{\mathbf{h}}$ is equal to the weight matrix of the family $A'=\{S_j(\mathbf{n})S_1^{-1}(\mathbf{n}):1\le j\le t\}$. In order to see that A' precedes A, simply observe what happens to the equivalence classes of A under $\sim$ upon multiplication by $S_1^{-1}(\mathbf{n})$. By Remark A5 (vii), every equivalence class other than that of S_1 is preserved. The equivalence class of S_1, on the other hand, can possibly be broken up into many class, but all these classes will be of weight less than $w(S_1)$. Hence A' contains the same number of equivalence classes as A for every weight greater than $w(S_1)$, and one less equivalence class of the weight $w(S_1)$. If follows that A', and hence $A_{\mathbf{h}}$, precedes A. We may therefore apply the induction hypothesis to $A_{\mathbf{h}}$ in (A.5).

Before doing so, however, we must keep in mind that the induction hypothesis only applies to families which contain no two expressions differing by a constant. The collection

$$\begin{aligned} A_{\mathbf{h}} =& \{S_j(\mathbf{n})S_1^{-1}(\mathbf{n}), S_j(\mathbf{n}+\mathbf{h})S_1^{-1}(\mathbf{n}) : 1 \le j \le t\} \\ =& \{S_j(\mathbf{n})S_1^{-1}(\mathbf{n}), S_j(\mathbf{n})S_j^{(2)}(\mathbf{n},\mathbf{h})S_j(\mathbf{h})S_1^{-1}(\mathbf{n}) : 1 \le j \le t\} \end{aligned}$$

need not have this property. Indeed, if S_j is a linear expression for some j then $S_j(\mathbf{n}+\mathbf{h}) = S_j(\mathbf{n})S_j(\mathbf{h})$ differs from $S_j(\mathbf{n})$ by the constant $S_j(\mathbf{h})$ (which is again the case after multiplication by S_1^{-1}). The question is, are these the only examples of two members of $A_{\mathbf{h}}$ differing by a constant?

The answer is "almost". Certainly $S_i(\mathbf{n})S_j^{-1}(\mathbf{n})$ is not constant for $i \neq j$ (by the properties ascribed to A), and for the same reason one may check that $S_i(\mathbf{n}+\mathbf{h})S_j^{-1}(\mathbf{n}+\mathbf{h})$ is never constant for $i \neq j$. (If m is the highest degree at which some polynomial appearing in S_j differs from the corresponding polynomial in S_i then $S_i(\mathbf{n}+\mathbf{h})$ and $S_j(\mathbf{n}+\mathbf{h})$ will display the same difference at this degree.) It is possible that for some values of $\mathbf{h}$ we could have $S_i(\mathbf{n})$ and $S_j(\mathbf{n}+\mathbf{h})$ differ by a constant for $i \neq j$, or for $i = j$ when S_i is not linear. However, by Remark A5 (vi) there exists a set $E \subset \mathbf{Z}^k$ of Banach lower density 1 such that for all $\mathbf{h} \in E$, $S_i(\mathbf{n})S_j^{-1}(\mathbf{n}+\mathbf{h})$ is not constant unless $i = j$ and S_i is a linear expression.

Let now $L \subset \{1, \cdots, t\}$ be the set of indices i such that S_i is linear, and put $N = \{1, \cdots, t\} \setminus L$. (A.5) can now be written as

$$\lim_{\substack{N_i - M_i \to \infty \\ 1 \le i \le k}} \frac{1}{\prod_{i=1}^k (N_i - M_i)} \sum_{\mathbf{n} \in \prod_{i=1}^k [M_i+1, N_i]} \int \prod_{j \in L} S_j(\mathbf{n})(f_j S_j(\mathbf{h}) f_j) \prod_{j \in N} S_j(\mathbf{n}) f_j S_j(\mathbf{n}+\mathbf{h}) f_j \, d\mu.$$

Multiply through by $S_1^{-1}(\mathbf{n})$ to see that this is equal to

$$\lim_{\substack{N_i - M_i \to \infty \\ 1 \le i \le k}} \frac{1}{\prod_{i=1}^k (N_i - M_i)} \sum_{\mathbf{n} \in \prod_{i=1}^k [M_i+1, N_i]} \int \prod_{j \in L} \left(S_j(\mathbf{n})S_1^{-1}(\mathbf{n})\right)(f_j S_j(\mathbf{h}) f_j) \prod_{j \in N} S_j(\mathbf{n})S_1^{-1}(\mathbf{n}) f_j S_j(\mathbf{n}+\mathbf{h})S_1^{-1}(\mathbf{n}) f_j \, d\mu.$$

For $\mathbf{h} \in E$ this is equal to (applying the induction hypothesis)

$$\Big(\prod_{j \in L} \int f_j S_j(\mathbf{h}) f_j \, d\mu\Big)\Big(\prod_{j \in N} \int f_j \, d\mu\Big)^2.$$

Since E is of Banach lower density 1, we therefore have shown that

$$\begin{aligned} & D-\lim_{\mathbf{h} \in \mathbf{Z}^k} \lim_{\substack{N_i - M_i \to \infty \\ 1 \le i \le k}} \frac{1}{\prod_{i=1}^k (N_i - M_i)} \sum_{\mathbf{n} \in \prod_{i=1}^k [M_i+1, N_i]} \langle x_{\mathbf{n}}, x_{\mathbf{n}+\mathbf{h}} \rangle \\ =& D-\lim_{\mathbf{h} \in \mathbf{Z}^k} \Big(\prod_{j \in L} \int f_j S_j(\mathbf{h}) f_j \, d\mu\Big)\Big(\prod_{j \in N} \int f_j \, d\mu\Big)^2. \end{aligned} \tag{A.6}$$

Recall that $\int f_{i_0}\,d\mu = 0$ for some i_0, $1 \le i_0 \le t$. If $i_0 \in N$ the limit (A.6) is clearly zero. If $i_0 \in L$, the limit is still zero since S_{i_0} is linear in that case and by Lemma A9

$$D-\lim_{\mathbf{h}\in\mathbf{Z}^k} \int f_{i_0} S_{i_0}(\mathbf{h}) f_{i_0}\,d\mu = 0.$$

So, in any event, the limit (A.6) is zero, which by Lemma A6 gives (A.4), as desired:

$$\lim_{\substack{N_i - M_i \to \infty \\ 1 \le i \le k}} \Big|\Big| \frac{1}{\prod_{i=1}^k (N_i - M_i)} \sum_{\mathbf{n}\in\prod_{i=1}^k [M_i+1, N_i]} \prod_{j=1}^t S_j(\mathbf{n}) f_j \Big|\Big| = 0$$

This holds provided $A = \{S_1, \cdots, S_t\}$ has no constant expressions. Recall that what we are really shooting for is (A.3):

$$UD-\lim_{\mathbf{n}\in\mathbf{Z}^k} \Big(\int \prod_{j=1}^t \Big(\prod_{i=1}^r T_i^{p_{i,j}(n_1,\cdots,n_k)} \Big) f_j\,d\mu \Big) = 0.$$

We need to establish this, where A is allowed to contain a constant expression. Utilizing just weak convergence in (A.4) we have

$$\lim_{\substack{N_i - M_i \to \infty \\ 1 \le i \le k}} \frac{1}{\prod_{i=1}^k (N_i - M_i)} \sum_{\mathbf{n}\in\prod_{i=1}^k [M_i+1, N_i]} \Big(\int \prod_{j=1}^t \Big(\prod_{i=1}^r T_i^{p_{i,j}(n_1,\cdots,n_k)} \Big) f_j\,d\mu \Big) = 0 \tag{A.7}$$

under these more general conditions. As a matter of fact, if we apply (A.4) to the product system $(X \times X, \mathcal{B}\otimes\mathcal{B}, \mu\times\mu, \{T_\mathbf{n} \times T_\mathbf{n}\}_{\mathbf{n}\in\mathbf{Z}^r})$, with f_i replaced by $f_i \otimes f_i$, $1 \le i \le t$, we get, utilizing just weak convergence,

$$\lim_{\substack{N_i - M_i \to \infty \\ 1 \le i \le k}} \frac{1}{\prod_{i=1}^k (N_i - M_i)} \sum_{\mathbf{n}\in\prod_{i=1}^k [M_i+1, N_i]} \Big(\int \prod_{j=1}^t \Big(\prod_{i=1}^r T_i^{p_{i,j}(n_1,\cdots,n_k)} \Big) f_j\,d\mu \Big)^2 = 0. \tag{A.8}$$

(A.7) and (A.8) together with Lemma A8 give us (A.3), as desired. This completes the induction step, so the theorem is proved.

□

Corollary A10. Let $k \in \mathbf{Z}$. The intersection in $\mathbf{Z}^k$ of finitely many sets of the form

$$\Big\{ (n_1, \cdots, n_k) \in \mathbf{Z}^k : \Big| \int \prod_{j=1}^t \Big(\prod_{i=1}^r T_i^{p_{i,j}(n_1,\cdots,n_k)} \Big) f_j\,d\mu \Big) - \prod_{j=1}^t \Big(\int f_j\,d\mu \Big| < \epsilon \Big\}, \tag{A.9}$$

where (for each of these sets independently) $(X, \mathcal{B}, \mu)$ is a probability space, $r \in \mathbf{N}$, $\{T_\mathbf{n}\}_{\mathbf{n}\in\mathbf{Z}^r}$ is a totally weakly mixing measure preserving $\mathbf{Z}^r$-action generated by $T_1, \cdots, T_r$, $t \in \mathbf{N}$ and $p_{i,j}(x_1, \cdots, x_k) \in \mathbf{Q}[x_1, \cdots, x_k]$ with $p_{i,j}(\mathbf{Z}^k) \subset \mathbf{Z}$, $1 \le i \le r$, $1 \le j \le t$ such that for any $1 \le j_1 \ne j_2 \le t$, the functions

$$(l_1, \cdots, l_k) \\ \to \big(p_{1,j_1}(l_1, \cdots, l_k) - p_{1,j_2}(l_1, \cdots, l_k), \cdots, p_{r,j_1}(l_1, \cdots, l_k) - p_{r,j_2}(l_1, \cdots, l_k) \big)$$

are not constant, and $f_1, f_2, \cdots, f_t \in L^\infty(X, \mathcal{B}, \mu)$, is of Banach lower density 1.

Proof. By Theorem A1 any expression of the form $(A.9)$ has Banach lower density 1. The result follows, therefore, from the fact that the intersection of finitely many sets of Banach lower density 1 has Banach lower density 1. (Equivalently, the union of finitely many sets of Banach upper density 0 has Banach upper density 0.) $\square$

REFERENCES

[B1] V. Bergelson, Weakly mixing PET, *Ergodic Theory and Dynamical Systems* **7** (1987), 337-349.

[B2] V. Bergelson, Ergodic Ramsey Theory–an Update, *Ergodic Theory of* $\mathbf{Z}^d$-*Actions*, M. Pollicott and K. Schmidt, eds., London Math. Soc. Lecture Notes Series **228**, Cambridge University Press (1996), 1-61.

[BFM] V. Bergelson, H. Furstenberg, and R. McCutcheon, IP-sets and polynomial recurrence, *Ergodic Theory and Dynamical Systems*, **16** (1996), 963-974.

[BH] V. Bergelson and N. Hindman, IP*-sets in product spaces, *Papers on General Topology and Applications*, S. Andima et al., eds., Annals of the New York Academy of Sciences **806**, New York Academy of Science, New York, NY, 1996.

[BL1] V. Bergelson and A. Leibman, Polynomial extensions of van der Waerden's and Szemerédi's theorem, *Journal of AMS* **9** (1996), 725-753.

[BL2] V. Bergelson and A. Leibman, Set-polynomials and polynomial extension of Hales-Jewett theorem. *Annals of Mathematics.* To appear.

[BM] V. Bergelson and R. McCutcheon, Uniformity in polynomial Szemerédi theorem, *Ergodic Theory of* $\mathbf{Z}^d$-*Actions*, M. Pollicott and K. Schmidt, eds, London Math. Soc. Lecture Notes Series **228**, Cambridge University Press (1996) 273-296.

[F1] H. Furstenberg, Ergodic behavior of diagonal measures and a theorem of Szemerédi on arithmetic progressions, *J. d'Analyse Math.* **31** (1977), 204-256.

[F2] H. Furstenberg, *Recurrence in Ergodic Theory and Combinatorial Number Theory*, Princeton University Press, 1981.

[FK1] H. Furstenberg and Y. Katznelson, An ergodic Szemerédi theorem for commuting transformations, *J. d'Analyse Math.* **34** (1978) 275-291.

[FK2] H. Furstenberg and Y. Katznelson, An ergodic Szemerédi theorem for IP-systems and combinatorial theory, *J. d'Analyse Math.* **45** (1985) 117-168.

[FK3] H. Furstenberg and Y. Katznelson, A density version of the Hales-Jewett theorem, *J. d'Analyse Math.* **57** (1991) 64-119.

[FKO] H. Furstenberg, Y. Katznelson and D. Ornstein, The ergodic theoretical proof of Szemerédi's theorem, *Bull. Amer. Math. Soc.* **7** (1982), 527-552.

[FW] H. Furstenberg and B. Weiss, The finite multipliers of infinite ergodic transformations, *Structure of Attractors in Dynamical Systems*, Springer Lecture Notes **668**, Springer (1978), 61-85.

[H] N. Hindman, Finite sums from sequences within cells of a partition of $\mathbf{N}$, *J. Combinatorial Theory* (Series A) **17** (1974), 1-11.

[Ha] P. Halmos, *Lectures in Ergodic Theory*, Chelsea, New York, 1960.

[KM] T. Kamae and M. Mendes France, Van der Corput's difference theorem, *Israel Journal of Math*, **31** (1978), 335-342.

[M] K. Milliken, Ramsey's Theorem with sums or unions, *J. Combinatorial Theory* (Series A) **18** (1975) 276-290.

[R] H. L. Royden, *Real Analysis*, 3rd edition, Prentice-Hall, 1988.

[S] A. Sárközy, On difference sets of integers III, *Acta. Math. Acad. Sci. Hungar.*, **31** (1978) 125-149.

[SZ] E. Szemerédi, On sets of integers containing no k elements in arithmetic progression, *Acta. Arith.* **27** (1975) 199-245.

[T] A. Taylor, A canonical partition relation for finite subsets of ω, *J. Combinatorial Theory* (Series A) **21** (1976), 137-146.

[W] P. Walters, Some invariant σ-algebras for measure preserving transformations, *Trans. Amer. Math. Soc.* **163** (1972), 357-368.

Vitaly Bergelson
Department of Mathematics
The Ohio State University
Columbus, OH 43210
email: vitaly@math.ohio-state.edu

Randall McCutcheon
Department of Mathematics
University of Maryland
College Park, MD 20742
email: randall@math.umd.edu

INDEX OF NOTATION

INDEX

Editorial Information

To be published in the *Memoirs*, a paper must be correct, new, nontrivial, and significant. Further, it must be well written and of interest to a substantial number of mathematicians. Piecemeal results, such as an inconclusive step toward an unproved major theorem or a minor variation on a known result, are in general not acceptable for publication. *Transactions* Editors shall solicit and encourage publication of worthy papers. Papers appearing in *Memoirs* are generally longer than those appearing in *Transactions* with which it shares an editorial committee.

As of March 31, 2000, the backlog for this journal was approximately 7 volumes. This estimate is the result of dividing the number of manuscripts for this journal in the Providence office that have not yet gone to the printer on the above date by the average number of monographs per volume over the previous twelve months, reduced by the number of issues published in four months (the time necessary for preparing an issue for the printer). (There are 6 volumes per year, each containing at least 4 numbers.)

A Copyright Transfer Agreement is required before a paper will be published in this journal. By submitting a paper to this journal, authors certify that the manuscript has not been submitted to nor is it under consideration for publication by another journal, conference proceedings, or similar publication.

Information for Authors and Editors

Memoirs are printed by photo-offset from camera copy fully prepared by the author. This means that the finished book will look exactly like the copy submitted.

The paper must contain a *descriptive title* and an *abstract* that summarizes the article in language suitable for workers in the general field (algebra, analysis, etc.). The *descriptive title* should be short, but informative; useless or vague phrases such as "some remarks about" or "concerning" should be avoided. The *abstract* should be at least one complete sentence, and at most 300 words. Included with the footnotes to the paper, there should be the 2000 *Mathematics Subject Classification* representing the primary and secondary subjects of the article. This may be followed by a list of *key words and phrases* describing the subject matter of the article and taken from it. A list of the numbers may be found in the annual index of *Mathematical Reviews*, published with the December issue starting in 1990, as well as from the electronic service e-MATH [**telnet e-MATH.ams.org** (or **telnet 130.44.1.100**). Login and password are **e-math**]. For journal abbreviations used in bibliographies, see the list of serials on the web at `http://www.ams.org/msnhtml/serials-list/annser_frames.html`. When the manuscript is submitted, authors should supply the editor with electronic addresses if available. These will be printed after the postal address at the end of each article.

Electronically prepared papers. The AMS encourages submission of electronically prepared papers in $\mathcal{A}\mathcal{M}\mathcal{S}$-TEX or $\mathcal{A}\mathcal{M}\mathcal{S}$-LATEX. The Society has prepared author packages for each AMS publication. Author packages include instructions for preparing electronic papers, the *AMS Author Handbook*, samples, and a style file that generates the particular design specifications of that publication series for both $\mathcal{A}\mathcal{M}\mathcal{S}$-TEX and $\mathcal{A}\mathcal{M}\mathcal{S}$-LATEX.

Authors with FTP access may retrieve an author package from the Society's Internet node `e-MATH.ams.org` (130.44.1.100). For those without FTP access, the author package can be obtained free of charge by sending e-mail to pub@ams.org (Internet) or from the Publication Division, American Mathematical Society, P.O. Box 6248, Providence, RI 02940-6248. When requesting an author package, please specify AMS-TEX or AMS-LATEX, Macintosh or IBM (3.5) format, and the publication in which your paper will appear. Please be sure to include your complete mailing address.

Submission of electronic files. At the time of submission, the source file(s) should be sent to the Providence office (this includes any TEX source file, any graphics files, and the DVI or PostScript file).

Before sending the source file, be sure you have proofread your paper carefully. The files you send must be the EXACT files used to generate the proof copy that was accepted for publication. For all publications, authors are required to send a printed copy of their paper, which exactly matches the copy approved for publication, along with any graphics that will appear in the paper.

TEX files may be submitted by email, FTP, or on diskette. The DVI file(s) and PostScript files should be submitted only by FTP or on diskette unless they are encoded properly to submit through e-mail. (DVI files are binary and PostScript files tend to be very large.)

Files sent by electronic mail should be addressed to the Internet address pub-submit@ams.org. The subject line of the message should include the publication code to identify it as a Memoir. TEX source files, DVI files, and PostScript files can be transferred over the Internet by FTP to the Internet node e-math.ams.org (130.44.1.100).

Electronic graphics. Figures may be submitted to the AMS in an electronic format. The AMS recommends that graphics created electronically be saved in Encapsulated PostScript (EPS) format. This includes graphics originated via a graphics application as well as scanned photographs or other computer-generated images.

If the graphics package used does not support EPS output, the graphics file should be saved in one of the standard graphics formats—such as TIFF, PICT, GIF, etc.—rather than in an application-dependent format. Graphics files submitted in an application-dependent format are not likely to be used. No matter what method was used to produce the graphic, it is necessary to provide a paper copy to the AMS.

Authors using graphics packages for the creation of electronic art should also avoid the use of any lines thinner than 0.5 points in width. Many graphics packages allow the user to specify a "hairline" for a very thin line. Hairlines often look acceptable when proofed on a typical laser printer. However, when produced on a high-resolution laser imagesetter, hairlines become nearly invisible and will be lost entirely in the final printing process.

Screens should be set to values between 15% and 85%. Screens which fall outside of this range are too light or too dark to print correctly.

Any inquiries concerning a paper that has been accepted for publication should be sent directly to the Editorial Department, American Mathematical Society, P. O. Box 6248, Providence, RI 02940-6248.

Editors

Selected Titles in This Series

(*Continued from the front of this publication*)

For a complete list of titles in this series, visit the
AMS Bookstore at **www.ams.org/bookstore/**.